APR 1 4 2009

GUIDELINES FOR

Integrating Process Safety Management, Environment, Safety, Health, and Quality

This *Guidelines* is one of a series of publications available from the Center for Chemical Process Safety. A complete list of titles appears at the end of this book.

GUIDELINES FOR
___

Integrating Process Safety Management, Environment, Safety, Health, and Quality

INTER-
SCIENCE

Copyright © 1996
American Institute of Chemical Engineers
345 East 47th Street
New York, New York 10017

All rights reserved. No part of this publication may be reproduced, stored in a retrieval system, or transmitted in any form or by any means, electronic, mechanical, photocopying, recording, or otherwise without the prior permission of the copyright owner.

Library of Congress Cataloging-in Publication Data
Guidelines for process safety management, environment,
  safety, health, and quality
      p.   cm.
  Includes bibliographical references and index.
  ISBN 0-8169-0683-1
  1. Chemical engineering—Safety measures. I. American Institute
of Chemical Engineers. Center for Chemical Process Safety.
TP149.G8356   1996
680′ .068′4—dc20   96-31457
                                                        CIP

This book is available at a special discount when ordered in bulk quantities. For information, contact the Center for Chemical Process Safety of the American Institute of Chemical Engineers at the address shown above.

It is sincerely hoped that the information presented in this document will lead to an even more impressive safety record for the entire industry; however, neither the American Institute of Chemical Engineers, its consultants, CCPS Subcommittee members, their employers, their employers' officers and directors, nor Arthur D. Little, Inc. warrant or present, expressly or by implication, the correctness or accuracy of the content of the information presented in this document. As between (1) American Institute of Chemical Engineers, its consultants, CCPS Subcommittee members, their employers, their employers' officers and directors, and Arthur D. Little, Inc., and (2) the user of this document, the user accepts any legal liability or responsibility whatsoever for the consequence of its use or misuse.

Contents

Preface	ix
Acknowledgments	xi
Glossary and Acronyms	xiii

1. Introduction   1

1.1. The Need for Integration	1
1.2. Purpose of Guidelines	3
1.3. Scope of Guidelines	4
1.4. Approach Used in Guidelines	5
1.5. Use of ISO 9000 Standards	6
1.6. Exclusions to Scope	7
1.7. Intended Audience for Guidelines	9
References	10

2. Securing Support and Preparing for Implementation   11

2.1. The Need for Securing Support	11
2.2 Identifying Who Will Benefit from Integration	18
2.3. Prepare a Preliminary Simplified Plan	20
2.4. Management Processes	26
2.5. Make Sure to Cover All Potential Benefits and Concerns	33
2.6. Mission Statement and Goals	37
2.7. Define Scope of Work and Approach	38
2.8. Selecting Your Integration Team	40
2.9. Project Status	41
References	41

3. Assessment of Existing Management Systems — 49
3.1. The Need for Assessing Existing Management Systems — 49
3.2. Assess Likely Support or Opposition to Integration — 53
3.3. Inventory and Assess All PSM, ESH, and Quality Management Programs and Elements — 56
3.4. Mapping the Management Processes — 58
3.5. Redesigning the Management Systems — 65
3.6. Update the Implementation Plan — 67
References — 68
Attachment 3.1 Selected Slides from Executive Summary of the Assessment of Existing Systems for Xmple, Inc. — 69

4. Develop a Plan — 73
4.1. The Need for Developing a Plan — 73
4.2. Adjust the Preliminary Plan — 77
4.3. Implementation Strategy — 81
4.4. Update Benefits and Costs — 85
4.5. Recast the Plan — 86
Reference — 88
Attachment 4.1. Sample Plans/Project Descriptions — 88

5. Integration Framework — 91
5.1. The Need for Developing an Integration Framework — 91
5.2. Prioritization of Programs, Elements, and Processes for Installation — 92
5.3. Developing Integrated Systems — 96
5.4. Continuous Improvement — 97
5.5. Quality Management Tools — 100
5.6. Converting Informal Systems — 107
Reference — 108

6. Testing Implementation Approach — 109
6.1. The Need for Testing — 109
6.2. Selecting the Pilot Project — 109
6.3. Establish Success (and Failure) Criteria — 113
6.4. Communication — 115
6.5. Conducting the Pilot — 116
6.6. Identifying and Correcting Deficiencies in Integration Plan — 117
Reference — 118
Attachment 6.1. Sample Pilot Project Advance Communication — 118

7. Tracking Progress and Measuring Performance　121
　7.1. The Need for Tracking and Measurement　121
　7.2. Capture Early Successes　122
　7.3. Measures to Consider　125
　7.4. Selection and Timing of Measures　129
　7.5. Customer Feedback　131
　7.6. Improving Performance　131
　Attachment 7.1. Sample Monthly Report　132

8. Continuous Improvement　135
　8.1. The Need for Continuous Improvement　135
　8.2. Management Responsibility　137
　8.3. Auditing the Quality System　138
　8.4. Product Verification　139
　8.5. Nonconformity and Corrective Action　140
　8.6. Personnel (Training)　140
　8.7. Use of Statistical Methods　141

9. Other Quality Management Systems　143
　9.1. Introduction　143
　9.2. Total Quality Management　144
　9.3. Malcolm Baldridge National Quality Award　144
　9.4. European Quality Award　145
　9.5. Deming Quality System　145
　9.6. ISO 14001　146
　References　147

10. Summary　149
　10.1. Introduction　149
　10.2. Case Study　149
　10.3. Summary　160

Appendix A: Overview of Definitions from ISO 9004: Quality Management and Quality Systems Elements—Guidelines　161
Bibliography　167
Index　169

Preface

The Center for Chemical Process Safety (CCPS) has recognized since its inception that enhancements in chemical process technologies, taken alone, are not sufficient to prevent catastrophic events. It is obvious that successful chemical process management technologies need the commitment and participation of top management. Therefore, with the support from its advisory and managing boards, CCPS established a multifaceted program to address the need for technical management commitment and technical management systems to reduce the potential for exposures to the public and to the environment.

Through a series of publications, CCPS has addressed the requirements and implementation of process safety management (PSM) programs. These were covered initially in the brochure *A Challenge to Commitment* which was sent to over 1500 CEOs in the chemical industry. *Guidelines for the Technical Management of Chemical Process Safety* was then published and it expanded on the elements of PSM. The third publication *Plant Guidelines for Technical Management of Chemical Process Safety* provided detailed examples and programs that could be implemented at plant levels. The fourth publication *Guidelines for Implementing Process Safety Management Systems* described the design, development and installation of PSM systems. CCPS appreciates that process safety management has considerable overlap with other environmental, safety, and health programs and that there are opportunities to share resources between these programs.

Quality management approaches are being applied widely in the chemical industry. CCPS has recognized that quality management approaches can be used to integrate process safety management and environmental, safety and health programs. These Guidelines will provide a framework and examples for integration of management systems designed to achieve continuous improvement in process safety, environmental, safety and health management. This book guides managers through the process of designing and implementing integrated programs.

Acknowledgments

The Center for Chemical Process Safety (CCPS) wishes to thank all the members of the Technical Management Subcommittee who provided guidance in the preparation of these materials. CCPS also wishes to express its appreciation to members of the Technical Steering Committee for their advice and support.

The Technical Management Subcommittee included: Frank P. Ragonese (Mobil Oil Corporation), past chairman; Thomas V. Rodante (Caltex Services Corporation), present chairman; Sanford Schreiber (CCPS staff member); Thomas A. Selders (ARCO); Jeff M. Gunderson (Chevron Research & Technology Company); Jim Parnell (Dupont Company); John Dowbekin (Exxon Chemical Company); William N. Helmer (Hoechst Celanese Corporation); Gary F. Hagan (Lockheed-Martin); E. J. Ryczek (Merck & Company, Inc.); Wayne E. Scheimann (Nalco Chemical Company); C. Robert West (PPG Industries, Inc.); Stanley Anderson (Rohm and Haas Texas, Inc.); John F. Murphy (The Dow Chemical Company); Alfred W. Bickum (The Goodyear Tire & Rubber Company); Leslie A. Scher (W.R. Grace & Company).

Arthur D. Little, Inc., Cambridge, Massachusetts, was the contractor who prepared these Guidelines. The principal authors were Lisa M. Bendixen and David A. Webb.

We gratefully acknowledge the comments and suggestions submitted by the following companies and peer reviewers: John Hoffmeister (Lockheed Martin Energy Systems); Mark Eidson (Stone & Webster Engineering Corp.); Thomas Janicik (Solvay Polymers, Inc.).

Glossary and Acronyms

Both quality management and environment, safety and health have specialized vocabularies. In some instances, different practitioners use the same word or words but assume different meanings. In this glossary we explain our use of these phrases.

Customer in quality management terms is the next person in the management process; of necessity it includes the ultimate customer who benefits from the final product. Each person in the management process is supplied with information, equipment, product, or raw material which must be worked on and then passed to the next "customer" for further refinement. A supplier is someone who provides information, equipment, product, or raw material. Thus a management process consists of a series of suppliers and customers.

Environment, Safety, and Health in this book generally refers to all programs intended to protect the environment, employee,s and third parties from any harm as a result of an upset condition in the operation of a facility using, processing, handling, or storing hazardous chemicals. In particular these include occupational safety, industrial hygiene, and all environmental protection programs. In this publication ESH is considered to exclude PSM.

Fishbone Diagrams are "cause-and-effect diagrams" used in quality management to help describe all the activities that can influence the management process and its outcome. These diagrams show the relationship between different activities and how they are grouped around specific types of activity.

Management Process in quality management refers to the activities conducted by all those involved in delivering the ESH or PSM program. It is not the process of chemical manufacturing, or the automated handling of raw materials, intermediates, or final products.

Pareto Diagrams are used in quality management programs to reveal the pattern of variation in performance and any predominant tendency. The data are displayed in the form of a histogram.

Process Safety Management (PSM) is the application of management systems to identify, understand, and control chemical and manufacturing process hazards and to prevent process-related injuries and incidents.

Quality Management is any approach to developing and implementing management systems that results in management processes which focus on what the supplier must deliver to meet customer requirements at every step of the process.

Stakeholder is a person or group that has a legitimate interest in the facility. For a facility handling hazardous chemicals, examples could be employees, neighbors, the board of directors, shareholders, and the regulatory community.

Total Quality Management (TQM) is a widely used quality management program (see Quality Management)

Acronyms

Throughout this book various acronyms and terminologies are used. We have attempted to explain these at the first occurrence. However, as it is generally impossible to rediscover the first occurrence quickly, this section provides a single point of reference to the puzzled reader.

AIChE	American Institute of Chemical Engineers
CCPS	Center for Chemical Process Safety of the American Institute of Chemical Engineers
ESH	Environment, Safety and Health
ISO	International Organization for Standardization
PSM	Process Safety Management
QM	Quality Management
TQM	Total Quality Management

1

Introduction

Since its founding in 1985, the Center for Chemical Process Safety (CCPS) of the American Institute of Chemical Engineers (AIChE) has promoted the enhanced management of chemical process safety. The CCPS program has always recognized that good safety performance is achieved through a combination of technology and management excellence.

In many organizations, the management programs for process safety, environmental, safety, health, and quality have developed separately. Yet these programs have many similarities and common needs. In an era when resources are becoming more scarce, managers may apply additional pressure to merge these management systems in order to provide more efficient and effective management of these issues. CCPS has recognized that Quality Management approaches have been widely adopted in the industry and believes that using Quality Management to integrate Process Safety Management (PSM) and other Environmental, Safety, and Health (ESH) programs and elements will provide significant benefits.

1.1. The Need for Integration

In almost every region and country, regulations are being introduced that require formal PSM and ESH management programs. In the United States, OSHA has implemented the PSM rule, and the EPA is extending these requirements with the proposed Risk Management Program rule. SARA Title III already imposes environmental management standards on many operations. In the European Union, the Seveso Directive and its successors introduced the need to identify and assess hazards and are now requiring formal management programs. Several countries on the Pacific rim have introduced regulations modeled on those in the United States and Europe.

In addition to regulations, public, owner, and political pressures require ever-better safety and environmental performance. Yet, at the same time, every company needs to find ways to reduce all costs in order to stay competitive. In the face of these apparently conflicting pressures, companies are looking for new ways to manage PSM and ESH issues.

Many corporations have adopted Quality Management programs throughout their organization. So far, these have been applied largely to discrete ESH programs and elements, and not to overall systems. For example, Quality Management approaches may have been used to develop the Process Hazards Assessment or Management of Change elements of PSM. A good example of this in practice is Dow Chemical Company's ESH auditing program, where a single process has been developed for all aspects of PSM and ESH. Another example is the Westinghouse Electric Corporation Management of Change program. In only a very few instances has there been an attempt to use Quality Management to integrate all PSM and ESH programs and elements into a single management system.

In manufacturing activities, the cost of poor quality is obvious: off-specification material, rework costs, scrap, wasted raw material, energy costs, equipment downtime and so on. Although some aspects of poor ESH management are obvious (injuries, business interruption, litigation costs, fines, clean-up costs, waste disposal) others are less visible. These less visible costs include: inefficient use of PSM and ESH manpower, time spent investigating and explaining incidents, lost stock value following a major incident, cost of installing "end-of-pipe" solutions rather than "designing-in" solutions.

In this book we will show that many aspects of ESH and PSM management systems are similar, including: auditing, hazard identification, equipment integrity and chemical hazards data. These programs and elements are part of every PSM and ESH program. Some organizations have started integrating these programs and elements with the intent of continuing down this path until all PSM and ESH programs and elements are included. The next step is to develop one integrated management system covering all the programs and elements; this is the subject of these guidelines.

The reward for successful integration is reduced cost of operation and more effective programs. The lower cost of delivery is achieved by developing management processes with fewer steps and no duplication of effort. Program effectiveness improves by adopting best available practices during the redesign. Finally programs designed using Quality Management approaches re-

> **The Need for Integration**
> - Increasing and overlapping regulatory demands
>   —Documentary and record-keeping requirements
>   —Formal and demonstrable programs
>   —Improved performance (particularly in areas such as emissions standards)
> - Pressure to reduce cost of operation and at the same time improve performance
>   —To maintain and improve competitive position
>   —To avoid costs of poor performance
> - Pressure to continuously improve ESH performance and stop taking "continuous corrective action" by correcting the underlying systematic failure. Another aspect of this is the tendency to continuously rework the same issues over and over again. Well-designed management systems should prevent this.
> - Recognition that other business activities have benefited from integra-

spond faster, at less cost and more effectively to new demands. Permanent fixes are installed to address what were previously recurring problems; "continuous corrective action" is eliminated or minimized. This is most clearly seen in areas such as inventory control and information systems.

1.2. Purpose of Guidelines

These guidelines present a process through which your organization could develop an integrated Process Safety, Environmental, Safety, and Health management system. This process is based on Quality Management approaches. Quality Management approaches are now widespread and most organizations have adopted a standard system, such as Total Quality Management or ISO 9000. The approach described in these guidelines uses the existing Quality Management and PSM and ESH expertise within your organization to develop an integrated system. This approach will provide a management system that is consistent with your company's culture and management style.

> **Purpose of Guidelines**
> - Presents an effective process for integrating PSM and ESH systems into one overall Quality Management system.
> - Demonstrates that this integration improves efficiency and reduces costs.

There are already a few companies that have wholly or partially integrated Quality Management into their ESH functions; the following examples illustrate the savings achieved by several of these organizations.

Examples of Cost Savings Achieved Using Quality Management Approaches

Company	Program	Savings Estimate
Xerox Corporation (Ref. 1)	Environmental Leadership Program	$100+ million/year
3M Corporation (Ref.1)	Pollution Prevention Pays	cumulative first-year savings of $506 million, 1975–1989
Unocal Chemicals (Ref. 3)	Safety Improvement Process	10% reduction in recordable incidents in first year
Tennant Company (Ref. 2)	Integrating Quality and Hazard Management	roughly 60% reduction in both injuries and defects

In addition each of these companies reports large reductions in waste and effluents.

1.3 Scope of Guidelines

These guidelines will show how a Quality Management system can be used to integrate PSM and ESH programs and elements. However, it is not our intention to introduce or describe Quality Management systems; other publications have addressed this comprehensively. You and your team should familiarize yourself with your company's Quality Management program by reading internal company publications or consulting standard references. A short and partial list of references is provided below. Our intention is to show you how you can extend your company's existing Quality Management system to provide an integration framework for PSM and ESH.

If your organization does not have a Quality Management system, you can still apply many of the concepts described in these guidelines. However, you will still need to identify a pervasive, consistent management system within which to conduct the integration.

> **Quality Management System References**
>
> Karl Albrecht and Lawrence J. Bradford, *The Service Advantage: How to Identify and Fulfill Customer Needs*, Dow Jones-Irwin, Homewood, Illinois, 1990.
>
> Frank Caropreso (ed.), *Making Total Quality Happen*, Report No. 937, The Conference Board, Inc., New York, N.Y., 1990.
>
> Chemical Manufacturers Association, *Questions of Quality, Integrating Process Safety and Total Quality: A Roadmap, Toolguide & Toolbox*, 1995.
>
> Joseph M. Juran, *Managerial Breakthrough: A New Concept of the Manager's Job*, McGraw-Hill Book Co., New York, N.Y., 1964.
>
> Edward J. Kane, "IBM's quality focus on the business process," *Quality Progress*, April 1986.
>
> William Scherkenbach, *The Deming Route to Quality and Productivity: Road Maps and Roadblocks*, CeePress Books, George Washington University, Washington, D.C., 1986.
>
> Peter R. Scholtes and Heero Hacquebord, "Beginning the quality transformation, part I; and 6 strategies for beginning the quality transformation, part II," *Quality Progress*, July-August 1988.

1.4. Approach Used in Guidelines

We have broadly broken the guidelines into segments that:

(a) Demonstrate the benefits to be gained by integrating PSM and ESH using your organization's Quality Management program in order to obtain management commitment to the proposed integration

(b) Demonstrate how PSM and ESH can be effectively integrated using Quality Management (through text and examples) to convince management of the viability of your proposal

(c) Show how you can build on and merge existing PSM and ESH management systems during the development of the integrated program. This will minimize the need to "reinvent-the-wheel" and maximize the probability of a successful integration

(d) Show how you can identify where common management activities can be combined and illustrate how these can provide a single overall management system

(e) Provide selected examples of integrated PSM/ESH/Quality Management systems and identify the quality tools you can use for integration

The guidelines assume that all PSM and ESH programs are to be integrated. In some organizations, only one program or element, such as training, may be intended to be integrated—with other programs or elements possibly following at a future date. In such cases, it will be necessary to carefully examine overlaps and interfaces with other programs and elements if the maximum benefits from integration are to be achieved. As much of the design as possible should be for a totally integrated system. The implementation can then be more piecemeal, but still be a part of the overall plan. Otherwise, each individual program or element may be optimized, but the overall system will not be.

1.5. Use of ISO 9000 Standards

The ISO 9000 series of standards has been used for illustrative purposes throughout this book. However, all the leading Quality Management systems have similar structures so you will find the concepts familiar regardless of which system your company has adopted. Chapter 9 summarizes the principal differences and similarities between ISO 9000 and other commonly used Quality Management systems.

The ISO 9000 series has been chosen because it is the most widely used Quality Management system and is a recognized international standard. The use of ISO 9000 in this book is not an endorsement of ISO 9000 over other systems. Rather, it is a practical decision based on the need to select one system to consistently illustrate the ideas contained in the book.

The ISO 9000 series consists of four standards:

- ISO 9001 Model for quality assurance in design/development, production, installation, and servicing
- ISO 9002 Model for quality assurance in production and installation
- ISO 9003 Model for quality assurance in final inspection and test
- ISO 9004 Quality management and quality system elements— Guidelines

In this book we have selected ISO 9001 as the model system and used ISO 9004 to identify the management process and quality system elements. These two standards are typically those used in the process industries. Although these guidelines are focusing on ISO 9001 and ISO 9004, your organization may use ISO 9002. ISO 9002 focuses on production and installation quality; it excludes design/procurement and servicing which is covered in ISO 9001. In the chemical industry it is unlikely that ISO 9003 would be used The scope of each of the ISO 9000 standards is summarized in Exhibit 1-1.

The ISO 14000 series of standards, *Environmental Management Standard*, has been established to specifically address environmental management. ISO 14000 is based on, and will therefore complement, ISO 9000 and will help identify specific environmental components for integration. If you chose to use ISO 14000, nothing in these guidelines will be invalidated.

ISO 14000 has been included in Exhibit 1-1 to illustrate its commonality with the ISO 9000 series.

1.6. Exclusions to Scope

The scope of these guidelines does not include advice on the development, testing or implementation of PSM or ESH programs and elements. The guidelines are focused on integrating existing programs and elements. There are several texts that address development, testing, and implementation, including the CCPS publications *Guidelines for Technical Management of Chemical Process Safety, Plant Guidelines for Technical Management of Chemical Process Safety* and *Guidelines for Implementing Process Safety Management Systems.*

Companies that have no quality system or that have just started implementing a quality system should be aware that the guidelines do not show how to select and implement quality systems. In the latter case, the book will still be useful in helping combine and streamline management systems. In addition, the comparison of different Quality Management systems in Chapter 9 may provide some useful input to Quality Management system selection.

1.7. Intended Audience for Guidelines

The intended audience for these guidelines includes those responsible for developing and implementing PSM and ESH management systems which could benefit from incorporation into a Quality Management system. This

ISO 9004 quality system elements/subelements	ISO 9001	ISO 9002	ISO 9003	ISO 14001
Management responsibility	●	⊛	○	√
Quality system principles	●	●	⊛	√
Auditing the quality system (internal)	●	⊛		√
Economics - Quality-related cost considerations				
Quality in marketing (contract review)	●	●		√
Quality in specification and design (design control)	●			√
Quality in procurement (purchasing)	●	●		√
Quality in production (process control)	●	●		√
Control of production	●	●		√
Material control and traceability (product identification and traceability)	●	●	⊛	
Control of verification status (inspection and test status)	●	●	⊛	
Product verification (inspection and testing)	●	●	⊛	√
Control of measuring and test equipment (inspection, measuring and test equipment)	●	●	⊛	√
Nonconformity (control of nonconforming product)	●	●	⊛	√
Corrective action	●	●		√
Handling and post-production functions (handling, storage, packaging and delivery)	●	●	⊛	√
After-sales servicing	●		⊛	√
Quality documentation and records (document control)	●	●	⊛	√
Quality records	●	●	⊛	√
Personnel (training)	●	⊛	○	√
Product safety and liability				
Use of statistical methods (statistical techniques)	●	●	⊛	
Purchaser supplied product	●	●		√

● Full requirement   ⊛ Less stringent than ISO 9001   ○ Less stringent than ISO 9002

Table adapted from Annex to ISO 9000 and Table 2 in ISO/DIS 14001

Exhibit 1-1. Comparison of ISO 9000 system requirements

> **Intended Audience**
> - Staff responsible for PSM and ESH programs
> - Staff at small, medium or large facilities handling hazardous chemicals
> - Locations with a Quality Management system

group includes both management and technical staff. We recommend that all managerial and technical staff who will be affected by the proposed integration should be given the opportunity to read these guidelines.

Although in large organizations, some integration efforts will be driven by corporate or divisional initiatives we expect that these guidelines will be equally useful at the facility level. Our target reader is most likely to be working at a large, medium or small facility handling hazardous chemicals and will already have a Quality Management system in place for other management processes.

These guidelines will be most useful to those facilities that have a requirement for a process safety management program. Within the United States this would include all facilities falling under the requirements of OSHA's Process Safety Management rule or the EPA's Risk Management Program rule. However, local regulations and corporate policies are resulting in the widespread adoption of process safety management programs and elements worldwide.

Finally, there will be the readers who are simply expanding their knowledge of PSM and ESH management. We believe that this group will also benefit from the ideas and examples contained in these guidelines.

In reading these guidelines, "you" refers not only to the reader, but also to the entire integration team. Thus, activities and requirements are not intended to be carried out by one person, but rather by one or more teams as described in Section 2.8 and elsewhere.

References

1. Abhay K. Bhushan, "Economic Incentives for Total Quality Environmental Management," IEEE, 1993.

2. Thomas J. Smith and Thomas L. Larson, "Integrating Quality Management and Hazard Management: A Behavioral Cybernetic Perspective," *Proceedings of the Human Factors Society 35th Annual Meeting*, 1991.
3. Stephen G. Minter, "Quality and Safety: Unocal's Winning Combination," *Occupational Hazards*, October 1991.
4. John F. Murphy, "Dow Chemical Company's Consolidated Audit," *Proceedings of AIChE 1992 Loss Prevention Symposium*.
5. Griff Holmes and William Leslie, *Management of Change and Total Quality Management Programs*, Westinghouse Electric Corporation, 1993.

2

Securing Support and Preparing for Implementation

2.1. The Need for Securing Support

The task of integrating PSM and ESH will be far easier and more likely to succeed if you have management support. Of course, commitment will be forthcoming only if managers believe the benefits outweigh the costs of integration. These costs include personnel costs, external consultants, other resources, disruption of existing PSM and ESH management and disruption of staff. It's not enough for management to commit to the process by saying: "I believe that integration will make our job easier." The commitment should be concrete: "I believe that integration will make our job easier and I am committing the following personnel and resources."

◆Hint
Obtain top-level management support and use this to leverage broad-based support for the integration of PSM and ESH.

It is also important to obtain broad-based commitment, because it is not possible to integrate these programs in a piecemeal fashion. However, implementation may be conducted step-by-step as a practical necessity. This means that top management must be committed as well as facility level staff. Top-level support helps makes sure that managers who will be involved in implementation will give priority to the integration. If integration is low on the list of company priorities, the integration effort will be starved of resources, will fail to meet targets and will most probably collapse.

Integration of PSM and ESH could be achieved without adopting Quality Management approaches. However, Quality Management approaches ensure that all issues are considered and represent a well-proven and widely used set of methodologies for designing effective management processes. But, unless your company has already introduced Quality Management approaches and used these successfully in other initiatives, you may want to consider delaying integration or using a different approach.

◆**Hint**
Read Chapter 2 of the CCPS publication *Guidelines for Implementing Process Safety Management Systems*, 1993, for guidance on obtaining buy-in. The principles described for obtaining support and commitment for PSM are equally applicable to integration of PSM and ESH.

The overall topic of obtaining support for any new initiative is developed in detail in the CCPS book *Guidelines for Implementing Process Safety Management Systems*, 1993, Chapter 2—Get Management Commitment. The following paragraphs summarize the approach detailed in those Guidelines. As you read these paragraphs, you should keep in mind the following issues:

- Who will benefit from improved systems and what will these benefits be?
- What will the final system look like?
- How will it differ from the existing system?
- How will the change be achieved?

A typical project has several phases (illustrated below). This chapter on securing support concentrates on the conceptual design and transition to the detailed design stage. You should remember that the team will need to secure support from different people at other stages of the project as well.

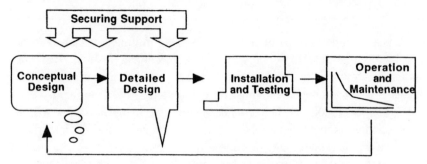

2.1. The Need for Securing Support

Preparation

- **Learn from successes and failures of comparable internal projects.** Informally review and consider major initiatives or campaigns (for example, PSM and Responsible Care®, or other quality management initiatives) that have succeeded in your company—along with those that have failed. What made them work? Incentives to participation? Relevance to job or quality of life? Other factors? And what about the failures? Lack of follow-through? Weak endorsement? Poor idea in the first place?

- **Understand who really makes the decision.** In deciding which manager(s) will be your primary sponsor(s), keep in mind that the terms "decentralized" and "centralized" can be misleading, because they are relative, not absolute. A company can be seen as decentralized because most operating decisions are made at the divisional level, working with broad corporate guidelines. A division which is part of the decentralized company may itself be a highly centralized entity, with key decisions coming from the top of the division and most staff resources based at division headquarters, not in the field. Of course, this could be the other way around with a highly centralized corporate entity and a decentralized division. Exhibit 2-1 provides some thoughts on the roles of different individuals in sponsoring and leading the project.

- **Communicate the right information.** Assessing your target audience's knowledge base and information needs is not always an exact science, nor does it need to be an exhaustive process. Rely on your own judgment and experience and seek out the opinions of others whose insights you trust. The key question you need to answer at this point is, "How much information does this person need from me to make an intelligent decision?"

    As a very general rule, it's better to presume less knowledge rather than more, because it helps assure that you cover all the bases. It is also easier to shorten a presentation (even while it's in progress) by skipping details than it is to add whole sections to it to fill gaps.

- **Identify all benefits.** Be creative and make sure you include benefits of all kinds in the omnibus list. While your selling effort should emphasize benefits, keep in mind that some legitimate benefits may be avoided negatives. For example, risk management is not generally designed to yield profits—but it will always help the company avoid a financial loss. Similarly, the best focus is probably on very tangible benefits, but don't ignore those that are less concrete. These might include improving employee morale through teamwork, or enhancing your company's image through favorable publicity about improved performance or your plans to strengthen PSM and ESH management.

- **Remember senior management is looking at the "big picture."** Most senior managers make decisions at a macro level. For example, a company's senior corporate management considers a $50 million acquisition to expand into a new product line, hoping to capture 15% of a $5 billion market over five years. Decisions such as this require assessment of cost/benefit to the corporation as a whole rather than in terms of impact on a specific process, facility or activity (for example, when a product or brand manager decides to redirect the marketing budget.) It is up to you to demonstrate your understanding of senior management's priorities and to frame your arguments in the context of the company and its industry standing. Present your rationale for integration in terms top management can appreciate, using language they can immediately grasp.

Selling the Concept

- **Be prepared; you may only get one shot at key managers.** If your selling opportunity includes an in-person presentation, consider conducting a dry run. In conducting a dry run, pay particular attention to the question-and-answer session: ask your colleagues to play devil's advocate and challenge your proposal. If you are preparing by yourself, review your proposal critically and try to anticipate the toughest possible questions.

- **Keep your audience and their agenda in mind at all times.** As you prepare your proposal, ask yourself at each stage if it answers two questions: "Why should we do this?" and "Why should we do it this way?" Answer: Integration will reduce costs and improve program effectiveness and Quality Management is a proven tool within the company and for improving ESH programs.

    Keep the presentation focused on the integration project; do not be drawn into discussions on particular strengths or weaknesses of the existing systems. If possible, illustrate expected improvements using existing measures of performance. Where performance measures do not exist, propose them, and propose to define the existing level of performance in order to measure future improvement.

    Finally, remember that in most instances your audience will have made their decision within five minutes. Show the benefits early and emphasize those that resonate with each particular audience.

- **Keep it a company-driven project.** Regardless of whether most of the groundwork is developed by employees or contractors, effective integration initiatives are company- or plant-driven, not imposed from outside. Effective implementation relies on a sense of investment among the people who ultimately will put theory into practice.

2.1. The Need for Securing Support

Designing the Concept

- **Describe the job you are asking senior management to do.** Try framing your expectations of senior management in terms of a job description. This may help you focus your own thinking and provide the basis for this section of your proposal. For example:
  —Manager Ex will ensure that appropriate resources are made available for the design, installation, and testing of an integrated PSM and ESH management system.
  —Manager Why will review progress on the integration project in monthly project meetings.
  —Manager Zee is responsible for ensuring the timely completion of the project within budget.

- **Make sure your proposal fits the corporate style and culture.** Make sure your expectations for senior management are consistent with your company's operating style. For example, a CEO who is accustomed to delegating most decisions is unlikely to accept a role that seems to take away divisional authority, while a facility manager with little functional autonomy will probably be leery of taking a highly visible role without approval from a superior. Of course, the CEO retains overall accountability for environmental and safety performance—that cannot be delegated!

- **Differentiate between goals and objectives.** When setting goals and objectives, it is important to understand the difference between them. The relationship between broad goals and specific objectives is comparable to the difference between policy and procedures. Goals help establish what your company expects to achieve, while objectives delineate how those goals will be met. A goal might be something very general such as "It is our goal to integrate all PSM and ESH programs within a single management system." This goal will be achieved through a series of objectives such as "We will complete a pilot integration project at one facility within 18 months."

- **Consider both the company's short- and long-term objectives.** Make sure your company's short-term objectives (for example, updating chemical spill policy and procedures) are consistent with broader integration goals (for example, reduction in overall resources dedicated to PSM and ESH), and communicate them in ways that clearly encourage work habits and procedures that support integration. For example, your employees need to understand that a significant improvement in resources and efficiency is achieved by integration and not by improving each individual program and element.

- **Draw on existing material.** If your company is actively involved in industry initiatives such as Responsible Care® or PSM, a formal mission statement may have been prepared that could be helpful as a guide.

- **Carefully select a leader.** Who should lead the detailed design and installation/testing efforts? It is best to consider skills and experience, not job title. You want someone with good "people" skills, a solid understanding of PSM and ESH, experience with Quality Management, a track record of earning respect within the organization and good project management skills. You need someone to be both advocate and manager. This may be the original champion or manager for the conceptual design phase. However, the tasks of detailed design and installation/testing are much larger, and your company may have quite different expectations of the leader for these phases.
- **Don't forget your other responsibilities.** In considering human resource needs, don't forget to include your own role. You have probably championed part or all of the conceptual design effort, and your continued involvement will be invaluable, but it may not fit with your other responsibilities. If you believe your commitment to integration may temporarily displace other priorities, make sure to include provisions for them.
- **Make sure you include all costs.** In weighing the pros and cons of inside staff versus outside contractors, the most useful determinant is cost-effectiveness over time. It's easy to underestimate the true costs of using employee resources, since these expenses are "buried" within ongoing business. At the same time realize that the experience and expertise of contractors is lost when they leave, while internal staff will retain the experience as part of the corporate memory.
- **Prepare a preliminary plan.** Although you will not be in a position to develop a final project plan, you can prepare a preliminary one. You should provide estimates of the project schedule, when the benefits will begin to be seen, and manpower and cost estimates. You should pay particular attention to the timing of benefits. In particular, remember that benefits may be achieved during the review of existing systems where gaps or defficiencies in existing programs may be identified and corrected. Also, remember that benefits may be monetary or non-monetary and short or long-term in nature. An outline of a preliminary plan is provided at the end of this chapter.

Communicate

- **Use communications specialists to help develop key messages.** If your company has people responsible for internal and external communications, seek them out. Consider inviting a member of your company's corporate communications or public relations staff to join a discussion group. Once you have agreed on broad goals for integration, consider using

2.1. The Need for Securing Support

someone from your company's corporate communications or public relations staff to help you draft a formal mission statement. Mission statements that lack substance are easily dismissed as window-dressing. Once it's drafted, review your mission statement critically for the "fluff factor." This will help you: assure a well-written document and win communicators' support by involving them. A mission statement might be something such as:

> We will eliminate all duplication of effort within our PSM and ESH systems by integrating the elements, programs, and management processes wherever possible. We will achieve this integration by using Quality Management principles to design new systems. We will measure our improvement in terms of fewer resources, quicker response to new demands and improved PSM and ESH performance.

- **Communicate.** Find out where your company's formal communications originate. Who is in charge of putting out press releases, newsletters, and product information? Arrange to meet with those responsible to see how you might incorporate information about the integration project into these publications, and offer to provide sample texts.

- **Keep your material interesting.** In any broad communication material a questions-and-answer format, or an "Interview with the Chairman (or other senior sponsor)" can help present information and messages in a conversational manner that many readers find more interesting than a more static narrative. A sample interview-style article is given at the end of this chapter.

- **Provide support materials to key people.** Prepare a briefing document, drawn from the integration proposal and executive summary, that executives, colleagues, and staff can use in responding to internal or external enquiries or in crafting speeches and other presentations. This document is often framed in terms of providing answers to questions that employees or other interested parties may ask. Some sample questions and answers are provided at the end of this chapter.

- **Communicate regularly.** If you have established a schedule for project meetings, this can provide a tickler for updating internal communications. Otherwise, establish a timetable (for example, monthly) for providing your internal "sources" with current information about progress. These communications should keep your audience informed on the status of the project and the outlook, both short- and long-term, for new developments. For example, in your initial communications you might discuss the preliminary plan and keep this up-to-date in subsequent articles or reports.

Exhibit 2-1. Sponsorship Roles

Role	Responsibility	Function in centralized companies may be:	Function in decentralized companies may be:
Key Advocate:	Endorses initiative and supports goals, provides resources	CEO COO SVP Operations	Divisional President Product VP
Prime Mover:	Oversees initiative, sets goals, supervises resources	Facility Manager GM-ESH Product Manager	Facility Manager Facility Technical Staff
Champion:	Drives the initiative, makes it happen	Corporate Staff	Local Staff

Exhibit 2-1 illustrates the differing sponsorship roles you might expect from various senior managers. You should carefully consider the roles different managers play within your organization and decide what is the right role for you to ask them to play. Always remember that you may need to recognize individual interests, experiences, and biases as well as the function filled by that individual. For example, it may be that one of the senior managers previously held an ESH position; this individual may be the best prime mover regardless of the function he or she now fills.

The balance of this chapter will focus on issues specific to developing your proposal for integration, particularly the benefits that should be derived from the project.

2.2 Identifying Who Will Benefit from Integration

There will be a wide range of people who will be affected by and benefit from the integration of PSM and ESH, all of these must be identified so that their needs and concerns can be identified and addressed.

In Quality Management there are generally two groups identified: stakeholders, and customers and suppliers. Stakeholders are those who have an interest in and benefit from the safe operation of your facilities. Stakeholders can include:

2.2 Identifying Who Will Benefit from Integration

- Employees who expect to be provided a safe workplace—these include operators, maintenance workers, laboratory technicians and technical staff. Each of these groups has different needs.
- Contractors who expect to be provided with a safe working environment and understand their role in achieving this.
- Owners who don't want the value of the company to be harmed by poor performance.
- Local politicians and other community leaders who may welcome the employment your company brings, but be concerned that the risks may be too high.
- Neighbors who need to be certain that they will not be harmed by your operations.
- Managers who want easy-to-use and effective management systems that cover all PSM and ESH needs.
- Regulators who expect compliance with all regulations and standards.

In quality management terms, the customer is the next person in an activity or management process, including, of course, the traditional customer who receives the final product. Each person is supplied with information, equipment, product or raw material which must be worked on and then passed to the next customer for further refinement. Thus an activity or management process consists of a series of suppliers and customers. For PSM and ESH integration, customers include people such as ESH staff, operating staff and individuals to be trained.

A single management process may control all or part of several PSM and ESH programs and elements. For example, a training center will use the same overall management processes to design, deliver, and administer many different courses. Similarly, spill prevention is a requirement of PSM, occupational safety, industrial hygiene and environmental regulation. Each of these requirements exists for different reasons and may cover different materials or spills sizes, but the objective is the same—prevent spills. A single common process could control all of these.

◆Hint
- Make sure you identify everyone who may be affected by the program.
- Identify and describe how each group will benefit from integration.
- Identify and address any concerns they may have.
- Test your list and ideas informally during discussions with colleagues.

2.3. Prepare a Preliminary Simplified Plan

Before any plan can be developed you must first have a vision of what the fully integrated system might look like. It may also be helpful to contrast the existing arrangements for PSM and ESH with this vision of the future. In this way, the advantages of integration can be more completely understood.

In Exhibit 2-2, we include six different specific "risk" management programs and acknowledge that there may be other company-specific programs. Four of these are obvious: PSM, occupational safety, industrial hygiene and environment. Different companies may divide these programs and elements up differently or use different titles. For example, many companies have a discrete hazardous materials or distribution/transportation activity. The last two cover loss prevention, namely property and casualty insurance and workers compensation, and may be unexpected. Some companies are including these in their integration of PSM and ESH, as they

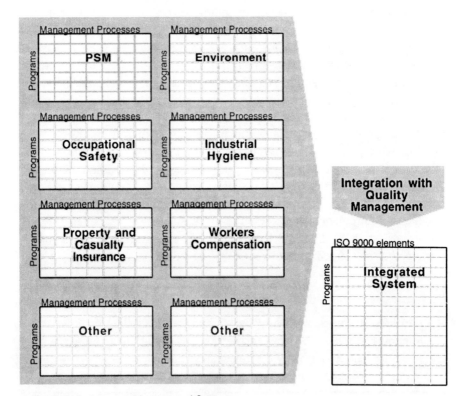

Exhibit 2-2. Vision of Integrated System

2.3. Prepare a Preliminary Simplified Plan

represent a significant cost (premiums and losses) that is closely related to PSM and ESH performance. Including these insurance activities has the benefit of providing a useful performance measure and locating control programs and costs within the same system. Other programs, such as product stewardship, might also be considered for integration. You should examine all potential candidates for integration within your organization. However, this guideline covers the integration of PSM and ESH and we will focus solely on these programs and elements.

In many companies, each of the PSM and ESH systems has developed independently with their own set of management processes, programs, and elements and resources. If these were integrated, the total number of programs would be significantly reduced by integrating common requirements. Moreover, a single set of management processes would cover all programs and elements. This streamlining would also reduce the staff resources required and provide a more flexible system capable of integrating change more quickly. The resulting system would be more responsive and better able to adapt to changing business and regulatory requirements.

In preparing the initial justification for integration, you will need to develop some supporting material; this process is illustrated in Exhibit 2-3.

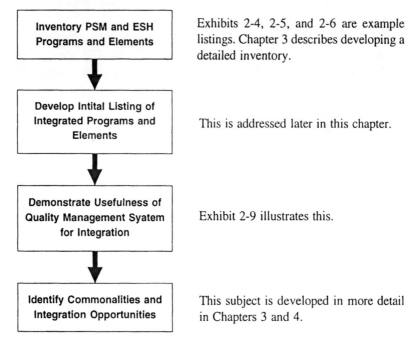

Exhibit 2-3. Developing the Initial Justification for Integration.

In preparing your proposal, the first step will be to inventory all the PSM and ESH programs and elements that exist, or should exist, within your company. To help with this, Exhibits 2-4, 2-5, and 2-6 provide lists of typical programs and elements. Although some of the titles vary, there is considerable overlap between these three exhibits. For example, they each have emergency response, hazards materials programs and work permit requirements.

♦Hint
In developing your list of PSM and ESH programs and elements, these sources may be useful:
- Training Manuals
- Audit Protocols
- Standard Operating Procedures

The PSM programs and elements presented in Exhibit 2-4 are drawn from CCPS books, API 750, the EPA Risk Management Program rule and OSHA's CFR 1910.119. It is impossible to develop a completely consistent list as the programs and elements in each of the programs and elements differ slightly in content and scope. It is important to note that PSM is actually a single program where many of the required management process elements are already defined. Thus PSM and Quality Management provide a basic structure on which an integrated PSM and ESH process may be built. As some PSM elements are regulated in the United States they are requirements that cannot be omitted.

There is a significant commonality in many PSM and ESH programs and elements which will make integration simpler. Simplistically, every PSM and ESH element or program can be described as being delivered by one or more of the following steps:

1. **Identify the hazards that must be controlled.** Without identifying the hazards it is difficult to decide whether a program is needed at all and impossible to design one that will control the particular hazards you face.
2. **Design a program to control those hazards.** Each program or element should control the actual hazards you have. It can be tempting to build something that addresses additional or greater hazards "to be on the safe side" and to guard against additional hazards in the future. This most often happens when you adopt a

2.3. Prepare a Preliminary Simplified Plan

Exhibit 2-4. Comparison of Different "Process Safety Management" Programs and Elements

	CCPS	OSHA PSM	EPA RMP	API RP 750	CMA Responsible Care®— Codes of Practice
Commitment and Accountability	✓				✓
Employee Participation		✓			
Management System	✓		✓		✓
Process Safety Information	✓	✓	✓	✓	✓
Process Hazards Analysis	✓	✓	✓	✓	✓
Hazard Assessment	✓		✓		
Operating Procedures/Safe Work Practices	✓	✓	✓	✓	✓
Training	✓	✓	✓	✓	✓
Contractors		✓			✓
Pre-Startup Review	✓	✓	✓	✓	✓
Maintenance/Mechanical Integrity	✓	✓	✓	✓	✓
Hot Work Permits	✓	✓		✓	
Management of Change	✓	✓	✓	✓	✓
Accident/Incident Investigation	✓	✓	✓	✓	✓
Emergency Planning and Response	✓	✓	✓	✓	✓
Audits	✓	✓	✓	✓	✓
Trade Secrets		✓			
Risk Management Plan			✓		
Registration			✓		

Note: For all these systems some of the "missing" elements and programs are contained within other elements and programs.

program from outside your facility without tailoring it to meet your needs. If your program exceeds your needs, there is a danger that employees will come to recognize this and begin to disregard particular aspects of the program. This erosion can eventually lead to the failure of the whole program.

3. **Install and test that program.** All too often a carefully designed program is improperly installed. There are many facilities that have excellent manuals describing their programs, but these bear no resemblance to what exists "in the field."

4. **Monitor the operation and effectiveness of the program for continuous improvement.** Over time your program will need to respond to issues such as new hazards or regulations, changes to management systems and design and operational changes. Only careful monitoring can keep the program up to date. This monitoring also checks that the program is being implemented as designed.

Each of these four steps can be a quality management process. In practice, other steps or substeps are also needed to achieve a successful program.

Exhibits 2-5 and 2-6 are modified sample lists of programs and elements for an actual company (here called Xmple, Inc.) drawn from the files of Arthur D. Little. These are in addition to the PSM requirements shown in Exhibit 2-4. You should develop similar lists for your company. These can usually be obtained from PSM and ESH manuals or the specialist staff who support the programs and elements. Be careful to make sure you understand the scope of each program, as the titles can sometimes be a little misleading. For example, spill response may cover only measures to be taken to stop further spillage, while containment and clean-up are covered under emergency response.

Even a quick examination of Exhibits 2-5 and 2-6 points out the considerable duplication that exists within Xmple, Inc. Not only are there separate management systems delivering their three programs (PSM, occupational safety and industrial hygiene, and environment) there are also several duplicate programs. For example, storage of hazardous chemicals and spill response are covered by all three systems. The lists are also incomplete; undoubtedly Xmple, Inc. implements all requirements of Responsible Care®, but it is listed only under occupational safety and industrial hygiene.

The next step is to develop a common list of programs and elements for PSM and ESH. This requires some understanding of each set of programs

2.3. Prepare a Preliminary Simplified Plan

Exhibit 2-5. Example List of Occupational Safety and Industrial Hygiene Programs and Elements for Xmple, Inc.

Industrial Hygiene	Occupational Safety
Industrial hygiene exposure characterization	Fire protection
Respiratory protection	Walking and working surfaces
Hearing conservation	Electrical system design
Ergonomics	Electrical safety-related work practices
Ionizing radiation	Control of hazardous energy
Nonionizing radiation	Confined space entry
Biological safety	Line breaking
Bloodborne pathogens	Machine guarding
Laboratory safety	Welding, cutting, and brazing
Substance-specific programs	Personal protective equipment
	Accident/injury record keeping
	Hazardous Waste Operations and Emergency Response (HAZWOPER)
	Fire prevention and emergency action plans
	Fire brigade
	Means of egress
	Flammable/combustible liquids—bulk storage
	Flammable/combustible liquids—portable storage
	Compressed gases
	Spray finishing
	ADA facility accommodations

and elements. You may wish to develop a "straw-man" proposal of your own or work with experts in each of the areas of PSM and ESH. However you proceed, we recommend that you first create a table similar to that illustrated in Exhibit 2-7. In Exhibit 2-7 we have attempted to collect comparable programs together under generic titles. You should select titles for these clusters that are meaningful within your organization, such as Information Requirements, Hazards Analysis and Work Permits. Looking at one of these examples, work permits, in more detail, you can see that psm included two categories of work permits, occupational safety has five and environment one. Exhibit 2-7 is incomplete; you should create a complete version for your organization.

Exhibit 2-6. Example Programs and Elements of Environmental Management for Xmple, Inc.

> Air pollution control
> Water pollution control
> Water supply management
> Solid waste management
> Hazardous materials management
> Soil and groundwater contamination
> Noise management
> Resource conservation
> Products and packaging
> Emergency preparedness
> Spill prevention
> Spill response
> Responsible Care (CMA) Code of Practice—Pollution Control

As you begin to work on Exhibit 2-7, it will quickly become apparent that each set of programs and elements has been created differently. The resolution of these differences will need input from the appropriate experts. For example, in Exhibit 2-7 we have shown that both occupational safety and environment have information requirements, yet these were not listed in the programs and elements shown in Exhibits 2-5 and 2-6. In preparing this table it is also important to make sure you understand what each element or program entails—names can be misleading. For example, human factors might also be ergonomics or operability, mechanical integrity might also be inspection and outreach might be risk communication or community involvement.

At this stage you should not be trying to develop a final list of integrated programs. This will come only after you have conducted a more detailed assessment of the existing programs. Keeping this limited objective in mind should cut short the inevitable debate that preparing such a list can generate.

2.4. Management Processes

All the programs and elements listed in Exhibits 2-4, 2-5, and 2-6 are delivered by management processes. Many organizations have at least three separate sets of management processes, each controlling one set of programs and elements. These are often almost identical. Integrating these programs

2.4. Management Processes

Exhibit 2-7. Example of Common List of PSM and ESH Programs and Elements

Clusters of PSM and ESH Programs and Elements	Focus of Current Programs and Elements		
	PSM	Occupational Safety & Industrial Hygiene	Environment
Information Requirements			
- Process Safety Information	X		
- Exposure Limits		X	
- Emissions Standards			X
Hazards Analysis			
- Process Hazards Analysis	X		
- Job Safety Analysis		X	
Work Permits			
- Hot Work Permit	X		
- Safe Work Practices	X		
- Confined Space Entry		X	
- Line Breaking		X	
- Machine Guarding		X	
- Welding, Cutting, Brazing		X	
- Personal Protective Equip.		X	
- Spill Prevention			X
Training	X	X	X
Emergency Response			
- Emergency Planning & Response	X		
- Fire Prevention and Emergency Action		X	
- HAZWOPER		X	
- Spill Response			

will result in better overall coordination of the programs, the natural evolution of integrated programs and the elimination of unnecessary overlap.

Before starting to think about what the new management processes might look like, you should make sure you have some idea how the existing processes function. ISO 9004 defines quality system elements. These are the requirements for the management processes that deliver a Quality Management system (see Exhibit 1-1). However, at this initial stage you do not want to develop a detailed description of the existing systems. The time for this is later, after initiation of detailed design when you will have the resources needed to undertake the scale of investigation needed. Rather, you need to have a general idea of how PSM and ESH are managed within your organization.

Fortunately, most PSM and ESH management processes have many similarities. They should each have a step which identifies new requirements and identifies the hazards that need to be controlled. This identification step is then followed by an assessment phase where decisions are made on whether the hazards are tolerable or new regulations are applicable to your operations. If a hazard is intolerable or the regulation applicable, the next step is to design a control program. Once designed, these programs must be installed, then tested and maintained. Generally each system in your organization will have some additional steps and side processes that are unique. You should develop simple flow diagrams (see Chapter 3 for some examples) to illustrate the processes in your organization.

Exhibit 2-8 is a generic process flow diagram for an integrated process. You should develop such a diagram for your organization. Make sure it covers any general internal control or policy requirements. Such a diagram is a powerful indication that integration is feasible. The features of a generic flow diagram are described in more detail below.

◆Hint

- If you cannot quickly identify the management process, it probably doesn't exist.
- Don't expend too much time trying to identify PSM and ESH management processes. They may not exist, be very informal or vary dramatically from department to department.
- It is quite likely that several other processes within your company, already "converted" to quality systems, also "didn't exist."

In most cases, your existing PSM and ESH management processes will already contain the basic elements of a quality management system—Planning, Doing, Acting, and Checking. When you construct the simple management flow diagrams for your organization, you should look for these four activities. If an activity is missing it may be due to an oversight, or it may not be part of your program.

In Exhibit 2-8 the need for new, or modified, control programs or elements could arise either because you have introduced a new process or facility, or modified an existing one. These changes may introduce new process hazards, new occupational safety or industrial hygiene issues or new environmental concerns. Any new hazards need to be assessed so that the company can decide whether they are tolerable or require new controls to

2.4. Management Processes

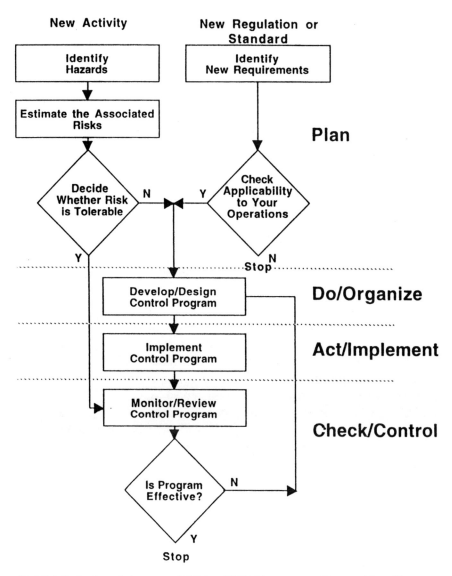

Exhibit 2-8. Sample Integrated PSM and ESH Management Process Flow Diagram

be kept at a tolerable level. For example, a new gasoline storage tank within an existing gasoline storage area built to current standards may need no control programs beyond those already in place for the other tanks. However, the introduction of a new chemical, not previously used at your installation, will probably need additional control measures. These measures could include Material Safety Data Sheets, additional personal protective equipment, handling procedures, additional testing and inspection requirements, additional emissions monitoring, and others.

New or modified regulations, industry standards or internal standards or policies can also impose new requirements on your existing programs and elements. You must decide whether they apply to your activities. Sometimes, you may be able to modify your operation to avoid such new requirements. If you do this, you will need to review these changes for any new hazards that may be introduced. A common example of this is the use of chlorine for water treatment. Some water treatment facilities found they fell under OSHA's PSM rule through the quantity of chlorine they stored on-site. They could avoid the requirements of PSM by reducing the size of the chlorine container or switching to alternative treatment chemicals. However, smaller containers increase the frequency and nature of refilling or replacement operations and new chemicals can affect the materials of construction. Such changes need to be assessed.

When you have decided that new or modified controls are required, you must design and develop them. This step may require physical modification to facilities, changes to operating procedures, additional monitoring activities or changes to operating conditions or safe operating parameters. Installing the new controls will require modification to written guides such as manuals and procedures, and training for all affected staff.

The new control systems must be monitored for effectiveness and your operations monitored to make sure you respond to any changes that bring your operations under the requirements of existing regulations and policies. The monitoring program is designed to determine the effectiveness of the controls. If these are found to be inadequate, they must be modified.

By preparing simplified management process flow diagrams for both existing and integrated systems, you can illustrate the improvements that will be achieved by integration. The most obvious improvement is the reduction in the number of separate management processes. Other improvements are the coordinated development of programs across more than one PSM or ESH program or element.

Once you have identified all the PSM and ESH programs and elements within your organization, you can develop a chart that shows the potential synergies with your Quality Management program. One axis of this table will be the programs and elements you identified and listed in Exhibit 2-7. The other axis will be the requirements of your Quality Management system.

Exhibit 2-9 illustrates the degree of overlap between PSM, ESH, and your Quality Management system. You should use this table to demonstrate

Exhibit 2-9. Example Matrix of PSM/ESH Programs and Elements and ISO 9000 Requirements

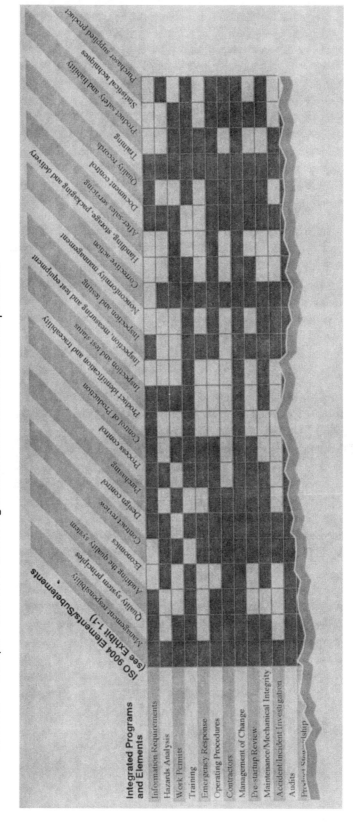

This is an example table. Each organization should have an equivalent but unique exhibit.

the likely efficacy of using a Quality Management approach to integrate PSM and ESH.

Each filled cell in Exhibit 2-9 shows where PSM or ESH programs and elements meet a particular ISO 9000 requirement. Exhibit 2-9 provides a powerful visual impression of the overlap between PSM, ESH, and Quality Management. Further, it is possible to quickly see where the greatest benefits may arise. By examining Exhibit 2-9 initial priorities for attention in our example might be:

- Training, operating procedures, management of change and audits are obvious first candidate programs and elements because these already have considerable overlap with the ISO 9000 requirements.
- Quality Management system priorities to consider are: management responsibility, quality system principles, auditing, contract review, design control, corrective action, document control, quality records, training, product safety and liability, and statistical techniques, because these are critical aspects of the Quality Management system where Xmple Inc. already has some systems in place.

Following the detailed review of existing systems, the relative strengths and weaknesses of each of the high priority areas can be assessed and priorities established for attention during the project.

Preparing your version of Exhibit 2-9 will provide a good indication of whether your PSM and ESH programs and elements will benefit from your Quality Management system. It will also help you identify where the greatest benefits will be obtained. Exhibit 2-9 could be used to represent either the existing system requirements within your organization, or your expectation of future requirements once an integrated system is in place. Which route you take will depend on your judgment of how to best present your proposal within your organization.

From Exhibit 2-9 it can be seen that there is a strong correlation between ISO 9000 and PSM and ESH programs and elements. However, the overlap is not complete. It should be remembered some PSM and ESH programs and elements are complete management process on their own (such as emergency planning), while others are one part of another process. For example, design reviews will generally be one part of a project design management process. PSM and ESH programs and elements that are stand-alone processes will have almost complete overlap with ISO 9000.

2.5. Make Sure to Cover All Potential Benefits and Concerns

Senior managers must consider a whole range of issues before deciding whether to support a new initiative. You must think as broadly as possible about the benefits that could be generated by integration. You don't want to leave out the one benefit that would have persuaded a key manager. We have already covered improvements in efficiency and effectiveness. Other possible benefits are:

- *Customer focus and satisfaction.* This requires that all of the customers of the processes and their interests be identified. Using a Quality Management approach ensures a deliberate effort to identify all customers and discover their expectations. The result will be that employees and other customers will experience improved satisfaction with PSM and ESH management processes.

- *Continuous Improvement.* Quality Management approaches require that continuous improvement be built into the processes. Staff responsible for implementing the programs will be encouraged to look for ways to improve both the efficiency of the processes and PSM and ESH performance. Generally, well designed Quality Management systems show significant improvement within a short time of start-up, followed by continuous steady improvement.

- *Written Standards.* Many existing programs may have been developed informally without written standards or expectations. A Quality Management approach requires that such standards be proposed, approved, and adopted. These standards form the basis for any program, they outline the objectives and they should be based on customer needs.

- *Measurement* must address end-of-pipe performance (through measures such as injury rates and gallons spilled), process efficiency and indicators of performance (through measures such as in service failure of equipment). Using indicators should make it possible to identify problems before they result in poor performance. Process efficiency measurement helps track implementation of the project and continuous improvement in the use of resources.

- *Statistical concepts* in many quality systems provide the basis for identifying the source of problems and directing improvement efforts toward these areas. Most of you will be familiar with the analysis of injury statistics to help decide where occupational safety resources should be applied. The usefulness of these is directly related to the measurement data available for analysis.

- *Problem solving* in quality management requires everyone involved in the process to be represented and be part of developing a solution. This, combined with quality management tools, ensures that the chosen solution is workable and that fixing one problem does not give rise to another.

- *Higher quality is always possible.* Improvement can be better end-of-pipe performance, a more efficient process or lower costs for the same (or better) performance, and is always possible in one of these dimensions.

- *Lower cost* refers to both the cost of the ESH services and the cost of ESH incidents (fewer and smaller). Quality management approaches enable management to improve both of these.

Most managers will have some concerns about integrating PSM and ESH or the use of quality management to achieve integration. It is vital to anticipate as many of these concerns as possible, some common concerns are described below along with some thoughts on how to defuse them.

- *Span of control* is most likely to be a concern among those reporting to senior management. However, if top level responsibility for PSM and ESH is divided among several managers, it can arise at this level as well. For example, occupational safety may fall under human resources, while environmental is within the legal department.

    Resolution: Most managers will be familiar with the changes in span of control when other processes were "converted" to Quality Management processes. It would be helpful if you could find out whether these changes were generally well received or just tolerated. You can consult colleagues and others who you know are sympathetic to the idea of integration.

    What needs to be understood is that the role of managers changes in quality management systems. The day-to-day process is driven by the staff involved in the process, rather than the manger, and these staff are likely to be in every department in the company. The role of the manager is to monitor performance, be a catalyst for improve-

2.5. Make Sure to Cover All Potential Benefits and Concerns 35

ment and make sure appropriate resources are available—all within his or her span of control.

- *Implementation costs* include staff resources needed to design and implement the new programs and the possibility that performance may suffer during the changeover to and familiarization with the new programs.

  Resolution: Your company will have some experience of these costs for other quality programs. You should review these and if costs were excessive, try to learn from these experiences and show how they can be avoided this time. Hopefully, there will have been no excessive costs and you will be able to use the review information to defuse this concern.

- *No reduction in cost will be achieved.* Managers facing a proposal to change a familiar system often express the concern that the proposal will not succeed in reducing costs.

  Resolution: Almost all Quality Management processes have fewer steps than the programs they replaced. Once again you should rely on experience with other programs inside your company. The Quality Teams will have developed process flow diagrams for the original and new programs. These will illustrate the improvement in process efficiency and resources that should be expected in PSM and ESH. But a manager's concern may be based on personal experience; probe for this and commit to investigate it for yourself. You may find that there were some extraordinary circumstances or that mistakes were made in the design and implementation. Report back on your findings and proposals for avoiding a similar problem.

- *Will all issues be covered* by the new processes? Most PSM and ESH programs evolved over a large number of years. Gaps were found and plugged. There is a natural concern that designing new processes will result in gaps that may expose the company to potential liabilities.

  Resolution: The development of Quality Management processes starts with a comprehensive review of existing systems and a careful review of the needs the processes must meet. Showing this process should help reassure the manager that all issues will be covered. You should also indicate what staff are likely to be involved in these processes and ask whether there are any others that should be included.

Personalizing the benefits of the new programs to each manager can be a persuasive tool. Once again, some research may help you find out which benefits are most likely to be compelling to particular managers. Some common benefits are described below:

- *Less time spent on ESH issues.* For most managers, PSM and ESH are only a small part of their total responsibilities. Furthermore, they probably have little training in PSM or ESH management, and so feel a little uncomfortable dealing with these issues. Quality Management processes are run by the staff involved in the process, with less reliance on management intervention. Managers should find themselves spending less time making the process work and more time on other issues or concentrating on the larger business needs of PSM and ESH.

- *More effective management of change.* PSM and ESH programs and elements must frequently be updated to reflect matters such as new regulations, changes in manufacturing processes and changes in corporate organization. Quality Management approaches result in processes that are generally simpler to update and can be consciously built to be adaptable to change.

- *Fewer processes to manage.* Once the integrated processes are in place, the total number of management processes and the number of steps within those processes will be reduced. Less management attention will be required to oversee these streamlined processes.

- *Continuous improvement* is designed in to quality management programs. You may be able to demonstrate this in other quality initiatives within your company. Also, in many companies one or more aspects of PSM or ESH performance have stagnated and shown little improvement. A quality approach may reinvigorate these programs.

- *Better measurement of performance.* A common frustration in PSM and ESH is that end-of-pipe measurement is all that is available and it is too late to correct a problem once the incident has occurred. Quality Management requires that we seek out in-process measures and leading indicators of performance that will warn of potential problems before they exhibit themselves as incidents.

◆Hint
You can find top managers' concerns from a number of sources:

- Recent corporate initiatives such as cost cutting or communications
- Annual reports, which usually review corporate priorities
- Internal communications from top management
- Your personal network of contacts

Make sure you review all of these; if your customers discover that you haven't read the Annual Report, your credibility may be ruined!

◆Hint
Remind managers of all the areas in which continuous improvement could be achieved:

- Improved PSM and ESH performance
- Lower cost of operating the PSM and ESH management systems
- Quicker response to changing regulatory or corporate requirements
- Better focus on stakeholder and customer needs

At the beginning of this chapter, we provided some general advice on how to gain management support for integration. We recommended that you identify champions in top management. Once you have conducted all the preparation outlined above, you should be in a position to identify specific candidates for each role. Your champions will most likely have one or more of the following characteristics:

- Good experience in or with Quality Management
- Impatience with existing PSM and ESH processes and their inefficiency
- Reputation for embracing change

2.6. Mission Statement and Goals

Successful Quality Management programs usually have clearly defined missions and objectives. Of the organizations that have committed to integration, we have found no single example that meets our needs. The following mission statement and goals have been adapted from several sources.

> ### Sample Integration Project Mission Statement
>
> We will integrate our PSM and ESH program based ISO 9000 approaches, consistent with current knowledge and regulation. Management of these issues continues to be a corporate priority, fully integrated into our business. Sound ESH management is a key factor of effective corporate management.
>
> We will conduct internal ESH reviews and measure our performance and will publish the results.
>
> We will share ESH data and information with our staff, customers, and suppliers.
>
> We consider ESH risk management to be part of the normal conduct of our business, and we will seek to identify and manage risk wherever it occurs.
>
> We will seek to establish efficient processes for managing ESH risk that can ensure comprehensive and consistent management of all our risks.
>
> It is our objective to:
> - Comply with all applicable laws, regulations, and industry practices
> - Support and contribute to research to improve ESH management and performance
> - Promote and encourage good business practices that result in improved ESH performance
> - Make all staff aware of ESH risks and the programs in place to manage them
> - Eliminate all injuries to our staff or third parties from our operations
> - Continuously reduce potentially harmful emissions and waste generation

2.7. Define Scope of Work and Approach

Before you decide on your overall approach, you will need to understand the approach to Quality Management and scope of PSM and ESH within your company.

The first step is to understand the Quality Management system within your company. Your company's existing programs may be formal or informal or a mixture of both. Characteristics of these are briefly described below.

- *Formal systems* have well-documented descriptions of how the system functions and records exist that demonstrate the systems are working.
- *Informal systems* may function well without good documentation. Small plants or small companies often have very good quality

2.7. Define Scope of Work and Approach

systems where procedures or work instructions are well documented but the overall system is not, because everyone "just knows" how it operates. However, this lack of formality is a weakness that can lead to system failures. The system is particularly vulnerable when staff leave or responsibilities are changed. Furthermore, if the plant or product is modified there may be no documented, fixed reference point from which to build or modify management processes.

Existing PSM and ESH programs and elements may also be formal or informal. It is important to make sure you uncover the complete scope of these systems.

You will need to undertake a series of interviews to understand existing PSM and ESH programs. You may also need to conduct some interviews to confirm aspects of your quality management system. However, if the Quality Management system is not well documented this may indicate a weakness, and you should reconsider the wisdom of moving forward with the proposed integration. It is essential to test how the processes are actually functioning, as this frequently differs from any description, formal or informal.

Carefully plan the selection of interview candidates. You should first consider what resources you have available—both in time and travel costs (if relevant). As a rule of thumb, it will take one hour to interview a manager with no special responsibility for PSM or ESH, two hours for specialist managers and staff and no more than 45 minutes for operators and technicians. You must also allow an equal amount of time to write up notes and review any material collected during the interview. You can now decide how many people you can afford to interview. In selecting candidates, you should consider the following:

- Interview representatives of staff at all levels in the organization in order to confirm whether the programs are actually functioning as designed.
- If possible, try to follow the management processes and interview staff responsible for every aspect of the process.
- Try and interview across a variety of departments and functions to ensure that your findings are not biased by weaknesses or strengths in particular departments.

Before starting the interview process, you should prepare an interview guide. You should focus on identifying the processes that are used to manage

PSM and ESH. Chapter 3 provides more detailed advice on developing and using an interview guide.

To a large extent, your overall approach to integration will be determined by the approach taken in other quality initiatives within your company. Unless there are strong reasons to adopt a different approach, it is best to follow this well-trodden path. Staff are already familiar with the process and will feel more comfortable with your project if it is implemented using proven methodologies. Most Quality Management development projects have common features, including the following:

- Teams such as quality improvement teams (QIT), may be used for implementation of all Quality Management projects. These teams are usually established for specific improvement projects and consist of staff involved in the management processes under review. For this project, PSM and ESH specialists, operations, maintenance, and engineering representatives would be members of the team.
- Each project has a champion and strategist who oversees the project, ensures that adequate resources are available and helps overcome any obstacles.
- A committee of interested managers and functional experts is established to review the proposals and monitor progress.
- Networks are established across project and functional lines to enhance learning.
- Adequate resources are made available for all phases of program development and implementation.
- An organizational structure is established for the project which is consistent with other quality management projects.
- Existing Quality Management processes outside of PSM or ESH are benchmarked to identify successful practices within your organization. This benchmarking may be extended outside your organization.
- A formal communication plan is established to keep everyone appropriately informed on progress, problems, and their resolution.

2.8. Selecting Your Integration Team

Your project team should include representatives from all major groups that will be affected by the integration. These groups will support the project more readily if they feel that someone on the project represents their interests. Everyone in the team should be familiar with your company's Quality Management program. Ideally they will have previously been part of a Quality Team. As a minimum, they should have been trained in your Quality Management system.

All group working sessions should be led by someone with a strong background in Quality Management and good facilitation skills. It is important that these individuals have the ability to draw out ideas from the team without imposing their own ideas. These facilitators need not be full-time members of the team, but are called upon when needed.

2.9. Project Status

At this point in the integration project you have obtained buy-in from senior management and have some thoughts on where the benefits will arise and what problems may be thrown in your path. The next steps are to firm up your analysis of the current status of PSM and ESH management and develop a more detailed plan and cost for the project. At the same time you will be developing better estimates of the benefits that will be derived from integration.

References

1. Center for Chemical Process Safety of the American Institute of Chemical Engineers, *Guidelines for Implementing Process Safety Management Systems*, 1993.
2. ISO 9000: 1987 (E)

Example of "Interview with the Chairman" for internal communications

Q *Why have you decided to integrate PSM and ESH now?*

A We have seen the benefits that redesigning other business processes have brought the company in terms of quicker and more appropriate response to change. I expect to see these same benefits in PSM and ESH.

Perhaps, even more important, is the stagnation in our safety and environmental performance in the last five years. Until five years ago, we had seen dramatic year on year improvement in every aspect of safety and environmental performance. This has slowed down in all areas and ceased altogether in occupational safety. I expect that these changes will invigorate all our programs. Our current safety and environmental performance costs us at least $30 million each year in avoidable costs such as capital damage and clean-up costs. If we could put a dollar value on injuries to our work-force, the savings would likely be ten times this amount.

Q *The basis for the integration will be ISO 9000; why this and not another quality management system?*

A As a company we have adopted ISO 9000 and today most of our staff are familiar with this system. Changing to a different system would cause confusion. We did consider adopting ISO 14000, the new environmental standard. However, we have no experience with ISO 14000, and I consider it prudent to stick with something we already know. The integration itself will introduce enough changes in the way we work. We may revisit this decision is a few years when some of our competitors have tried out ISO 14000!

Q *One reason for integration is reducing the resources needed to manage PSM and ESH; will there be any layoffs?*

A No, most of our PSM and ESH management is done by line managers who have many other responsibilities. I want an integrated management system that allows them to spend more time focused on other business issues, while at the same time improving PSM and ESH performance. I do expect that we will reduce our corporate and divisional staffs in these areas, but these staffs are already small and anyone whose post is eliminated will be redeployed elsewhere. We have been allocating some of our best technical and managerial talent to PSM and ESH in recent years. This is our opportunity to spread this talent around the company.

Example of "Interview with the Chairman" for internal communications

Q *Where do you expect the benefits of integration to be greatest?*

A Without doubt in our ability to adapt to ever-changing regulatory and business needs. For example, when OSHA introduced their PSM rule we had to build many of our systems from scratch, despite knowing that we were managing all the issues that were addressed in the rule. We now recognize that we already had a great deal of what was needed in other programs. If we already had an integrated system, PSM would have cost us less and would have been implemented more quickly. Another example was our decision to back integrate into the manufacture of monomer for our fibers business. I lost count of the number of different ESH programs we needed to upgrade. Once again we can now see that we duplicated a huge number of systems unnecessarily.

Q *Where will the greatest problems be?*

A We have been managing ESH for many years and have well established programs that, for all their inefficiency, have generally served us well. Some staff and managers will be reluctant to give up the established systems in favor of a new and untried system.

Q *What can we do to overcome these problems?*

A Firstly, it is important to recognize the benefits we have derived in the same circumstances for other programs. One example is our inventory management systems. In the old inventory management system, each department was responsible for holding the critical inventories they judged to be necessary. When we looked into this more closely, we not only found wide variations in the criteria for critical inventory, we also saw enormous duplications. I recall that we found that we had enough heat exchanger tubes to refit every exchanger on our olefins complex and enough alumina to refill every drier in the company the next day! I expect we will find equally dramatic opportunities for improvement in PSM and ESH. We have already conducted some preliminary analysis and these charts illustrate the large overlap between our different programs. (See Exhibit 2-9.)

Q *What resources will be needed for the integration effort?*

A Initially a team of five or six will develop a detailed proposal. We will then select a division for a pilot study. The pilot will have a team of about 12 people, half of whom will be full-time. As the integration is rolled out in other divisions, I expect a slightly lower level of effort, as we will have learned from our earlier experiences. We will also use an outside consultant to help us develop our plans and review implementation.

Q What role will you be playing in the integration effort?

A I have asked our Corporate VP for Operations to take day-to-day oversight of the work. I expect to review progress every month. I will make sure that the project has all the resources and management commitment needed to make it succeed. I have already planned for an annual reduction in accident and incident related costs of at least 10 percent each year. I have given my personal commitment to the board that we will achieve this.

Example of briefing paragraphs for support of internal and external communication

The Interview with the Chairman questions and answers form part of this document.

Q Are you changing the PSM and ESH management systems because of unsatisfactory performance?

A Yes, although our performance is one of the best in the industry, our failure to improve in the last few years is disappointing. However, this is only one of several reasons. We have seen the benefits of ISO 9000 and redesigned business processes in other parts of our business. It is now time to get these benefits into PSM and ESH. We believe that an ability to respond faster to new demands will provide a business advantage. We will be able to permit new facilities faster and respond to new regulations and standards more quickly and at lower cost than our competitors.

Q Will this project take away resources that might have been used to improve business performance elsewhere?

A No. We have a long-established project approval system which applies to all projects. We prioritize our investments based on the return they will bring to the company. In the case of safety and environmental projects they would receive priority treatment only if we believed our current performance fell short of the standards expected of us.

Q How do you set safety and environmental standards?

A In a number of ways:
　　—The absolute minimum standard is compliance with all applicable codes and regulations in each jurisdiction we operate in.
　　—Beyond that, we annually set emissions, injury, and incident rate targets. Our performance against these is publicly reported in our annual report. Each year we review performance and set new, more demanding, targets.

—It is our corporate policy to provide a safe work-place. We try to listen to our employees to understand what safety and environmental concerns they have, we believe we can improve in this area and this will be addressed in the redesigned system.

—At each of our facilities we have established a community outreach program. Through this we try to learn about any concerns our neighbors have about our performance.

—We monitor the performance of our competitors and seek to learn from their experiences.

All of these are integrated into our target setting process.

Q *Setting performance targets suggests that you believe accidents are inevitable. Do you intend to manage performance at a particular level?*

A No. As the Chairman said in a recent interview–and I will repeat now–it is our intention to continually improve performance. However, it is unrealistic to believe that we could achieve a zero accident or emission standard tomorrow. Even if it were theoretically achievable, the cost would be so large that we would bankrupt the company. We must maintain a balance between the benefits we achieve and the cost of achieving them. Let me also remind you that as a company we have made a commitment to abandon any business or technology where we consider the safety and environmental risks to be intolerable.

Q *The integration is designed solely to reduce the direct costs of the PSM and ESH programs. You will not actually improve performance at all, will you?*

A Without an improvement in performance, we would be throwing our money away. The direct cost savings alone would not justify this project. Only when we factor in improved safety and environmental performance does the cost of this project make any sense. We would not be going ahead unless we were confident that real savings are possible.

Q *This redesign, like all the others that have been implemented, will put the responsibility for safety and environment onto the shoulders of operators and technicians. Management will now to able to put the blame for poor performance elsewhere.*

A All managers will continue to be held accountable for the safety and environmental performance of their departments. Indeed, part of their annual performance review depends on meeting specific safety and environmental objectives. But, the processes will be redesigned so that they require as little management intervention as possible. Based on our experience in other processes, the workload will be reduced for everyone. I am sure there may be some exceptions to this, but in these cases we will make sure that other tasks are reassigned elsewhere. If anyone feels unable to carry out some of the new responsibilities, I expect this will be raised during our consultation process.

Q *Will the trades unions be consulted on these changes?*

A When we introduced ISO 9000 a few years ago we negotiated agreements with the relevant unions. These agreements anticipated that ISO 9000 would eventually be applied to all our management processes. However, as we have done throughout the implementation of ISO 9000, we will make sure union officials are fully informed on our proposals and the progress of the project. The trade union officials are the representatives of our most important stakeholder—our employees.

Q *What about regulators? They are used to seeing systems dedicated to their area of interest.*

A The new systems will continue to meet all the reporting and documentation needs of the regulators. But we will be reviewing our program with local inspectors as we begin work at each facility.

Q *How will we ensure that during the transition from the existing to the new systems all PSM and ESH issues continue to be managed?*

A This is a real concern that will need to be managed. Our experience with ISO 9000 has been good. One of the first tasks is to identify everything that is being and should be managed by the new system. In many instances, we have found gaps in the existing systems. Once we have identified all the issues we will set about designing the new processes and then put together our transition plan. While I cannot describe the details of this today, I am confident that we have a process that will address these issues.

Q *Who will fund the project: corporate, the divisions, or the facilities?*

A We absorb all costs at the business unit level, that is, the divisions. However, corporate is setting the agenda for this work. Each division has been asked to include the costs of integration into its current planning cycle. Where this impacts short-term planned performance, we will adjust the plans. However, from the next planning cycle onward the plans will include these costs—and the benefits of the new system. There will be one exception to this: the conceptual plan and pilot project will be developed and funded by corporate.

> **Q** *What is the timetable for integration?*
>
> **A** We expect to complete a pilot project by the middle of next year and begin work in the divisions within six months of the pilot project completion. We expect to have the new systems in place throughout the company within three years.

3

Assessment of Existing Management Systems

3.1. The Need for Assessing Existing Management Systems

Before starting on the design of any new management systems it is vital to understand what needs to be managed and how this is currently achieved. Without such an understanding the team may develop a system that is incomplete, fails to take advantage of existing systems or cannot function within your organization. The team may also spend time designing new systems when they might have adopted an existing system from within your organization.

Experience indicates that in a typical company there will be considerable overlap between and within the PSM and ESH and quality management systems. Your work to secure support (Chapter 2) will have already demonstrated many of these overlaps. The detailed assessment of the existing management systems described in this chapter will provide a comprehensive understanding of the overlaps.

It is also useful to identify any drivers that may support new integrated systems—or any factors that may inhibit integration. By identifying these the team will be able to take advantage of support and work around any problems.

◆**Hint**

Read Chapter 4 of the CCPS publication *Guidelines for Implementing Process Safety Management Systems* for guidance on assessing existing management systems. The principles for assessing PSM are equally applicable to integration.

This chapter describes an approach to assessing the existing management systems. This approach is based on quality management principles and is similar to that described in the CCPS *Guidelines for Implementing Process Safety Management Systems*, Chapter 4, Evaluate the Present Status. The following paragraphs summarize the approach outlined in those guidelines, as adapted to meet the needs of integration.

General Issues for Assessment of Existing Systems

- "Do it once = Do it right." Assessing the existing systems takes time and effort, but it will ensure that your solutions will hold up over time. In many companies, this heavy "front end" runs counter to a more action-oriented culture, and there is a strong temptation to fix the problem and get on with it. Keep in mind that you are designing an overall management system first. Detailed programs and elements will follow. Developing an understanding of existing systems and any likely obstacles to or motivators for integration will provide the right starting point for designing the new system.
- Remember that "home-grown" solutions are frequently better received (and thus more effectively applied) than totally new approaches that seem to be artificially created and imposed. The more transferable ideas you are able to identify and incorporate into your implementation plan, the greater the likelihood of its success.
- Higher-risk activities require more formal management systems. For example, a process using a highly toxic raw material will require more extensive management control than a process using only benign chemicals. You are more likely to find mature and well-developed formal systems in facilities with high risks.
- If you do not have an in-house audit function, it may be wise to consider investing in a professional ESH auditing course for one or more staff members, who can then share their knowledge with others.

Detailed Design Considerations

- The assessment must be designed to be as objective as possible. An assessment that lacks objectivity may yield questionable results. For example, someone who has been involved in the development or use of a management process is less likely to see weaknesses than an independent observer. When selecting your assessment team, make sure to include staff who will not take things for granted and will bring a fresh set of eyes and ears.
- The evaluation tools and techniques you use should be kept constant. This helps ensure consistency among evaluations at different locations and by different staff. Make a trial run using your set of tools to make sure that

3.1. The Need for Assessing Existing Management Systems

you have ironed out as many of the bugs as possible. Unless you are using a team experienced in the application of your tool-set, consider running a training session for everyone who will be conducting the assessments. In selecting your assessment tools, follow these principles:

— Use clearly defined assessment criteria to indicate present levels of performance. In this case the management systems are being compared with your quality management system and should be assessed against them. PSM and ESH programs and elements should be compared against a simple model (something like that described in Chapter 2, Section 2.3).

— Look only at what is relevant at each location. For example, if policy is set locally, look at it at each location. If it is set centrally, look at corporate policy making and limit local review to an examination of policy communication and interpretation. Also, don't spend time assessing programs and elements that are not relevant to a location. For example, don't look for toxic substance control programs at a location with no toxic materials on site.

— Don't spend time collecting information already collected elsewhere. For instance, if your company has a comprehensive environmental management audit that looks at all environmental programs and elements, rely on these audits for that category of information. Be aware; make sure to really understand what information is collected and whether it will meet your needs.

— Get buy-in from local managers. Cooperation of local management and staff will make the whole project much easier and the assessment phase will be the first working contact local staff will have with the project. It is important to set off on the right foot. Using assessment approaches, such as auditing, that are familiar to local staff and are conducted by staff they know of and respect help ensure buy-in.

- **Audits.** If you select auditing as your assessment approach, you should consider the following:

    — The CCPS book *"Guidelines for Auditing Process Safety Management Systems"* provides guidance on developing audit programs and conducting audits.

    — To improve objectivity, choose auditors with appropriate functional experience, but who work at different locations within the company.

    — If your company has an in-house audit function or standing team, try to enlist their support, either on a consulting basis or to conduct the assessment under your team's direction. At the very least, to avoid potential misunderstandings, make sure they know about the assessment.

    — The audit should gather data from all levels in the organization. Getting time with senior managers can be difficult and it is usually best to make appointments with them ahead of the assessment dates.

- **Surveys.** If using survey by questionnaire as your assessment approach, consider the following:
  - Survey questionnaires are completed by local staff. Each facility may interpret the questions differently if their intent is not clearly specified along with the question. You could include intent statements to clarify the questions, since you will have no basis for understanding differences between locations. You will not be able to easily check their accuracy.
  - Surveys to gather advance information on the facility, their organization, and general information on PSM and ESH management can save on-site time of your assessment team.
  - If you gather most of your information through surveys, it is well to conduct spot checks, perhaps through telephone interviews.
  - Consider designing a questionnaire with open-ended questions and invite the person filling it in to use examples to illustrate particular points.
  - If the questionnaire is to be filled in by a large number of people you will generally solicit more honest answers by assurances that the information is confidential and having the forms returned directly to your assessment team.
  - When designing the questionnaire, consult with people in your company's data processing department, who can help build in tabulation requirements on the front end. Make sure they understand answers will be qualitative as well as quantitative, and ask for advice on structuring the questionnaire to facilitate consolidation of results.
- The key to effective interviewing is to spend much more time listening than talking. Approach each interview with an open mind and with as few preconceived ideas as possible about the facility, process or individual you're interviewing. Remember that the goal in interviewing is not to fill out a form, but to elicit essential information that will help guide your system development.
- Regardless of how the assessment is conducted, your assessors should take detailed notes using a common format to capture the maximum amount of information in a consistent manner. In addition to recording quantitative information, assessors should record observations and opinions for future use. It is important, however, to make sure that these observations and opinions are clearly marked to distinguish them from information obtained from the interviewee.
- Encourage auditors and assessors to review and organize their notes as soon as possible after completing an interview or site visit. It's easy to lose track of insights and observations if too much time elapses, particularly when an assessment involves multiple sites.

3.2. Assess Likely Support or Opposition to Integration

Quality Management systems have demonstrated their effectiveness in providing efficient, flexible management processes. However, in many instances they have failed to deliver their promises in full and in some cases have failed completely. These failures have been ascribed to cultural obstacles that were not identified during the design process. In a recent survey of 350 senior executives Arthur D. Little, Inc. found that 68 percent of the companies reported unanticipated problems with their change process.

In most instances cultural problems arise because there is misalignment between the new requirements and factors that drive day-to-day behavior—*the unwritten rules*. One model for the staff motivation process, shown in Exhibit 3-1, illustrates how staff behavior is influenced by:

- Motivators—Factors that drive particular behaviors.
- Enablers—People or systems that can provide the motivators.
- Triggers—Events that can initiate this whole motivation process.

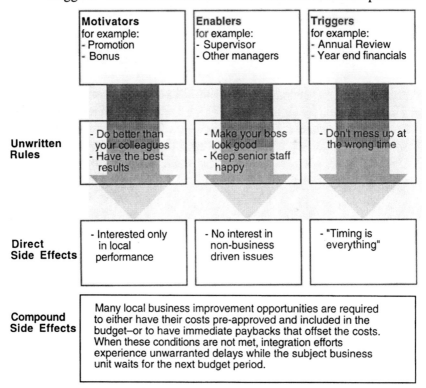

Exhibit 3-1. Influences on Staff Behavior

In the exhibit, one effect of a common set of unwritten rules has been illustrated. In this example, staff are driven to meet personal performance targets, watch issues that influence the performance of their bosses and make sure that results are not adversely impacted by new initiatives. For the integration initiative this could spell disaster and result in the delays referred to under compound side effects in Exhibit 3-1. However, it is possible to plan or design around this. One way would be to show that local performance will be improved, managers will more easily meet their PSM and ESH targets at lower cost, and all costs will be included in the relevant plan period. You might also help local managers negotiate an agreed change to their performance targets.

♦Hint
Look at previous quality management initiatives within your organization to see whether they have met, exceeded or fallen short of expectations. For those that failed to meet expectations, identify the problems that led to the shortfall and make sure to address these in your plan.

Consider including an unwritten rules assessment in the review of existing systems. This can be done either as a stand-alone activity or integrated into the general assessment. Either way you will need to familiarize yourself with unwritten rules assessments (Reference 2). The interview or survey questions should be designed to draw out the Motivators, Enablers, and Triggers that will be important to the success of your project. Exhibit 3-2 illustrates one way to think about designing these questions. Design questions that identify the "who, what, and when" that influence staff behavior. Aso try to identify examples of where these factors have influenced staff behaviors, either beneficially or negatively. Analyze your findings to identify how you can take advantage of unwritten rules and where you need to plan or design around them. One caution: any changes you introduce may change the unwritten rules themselves!

Quality Management systems can fail for reasons other than cultural ones. One of the most common failings is building Quality Management systems alongside existing programs. ISO 9000 has been a particular victim of this fault. Many companies have adopted ISO 9000 to attain ISO 9000 certification and have built systems solely to achieve this end. As a result, the new quality-designed systems do not replace existing systems. As these additional systems provide no direct business benefit, they may be regarded

3.2. Assess Likely Support or Opposition to Integration

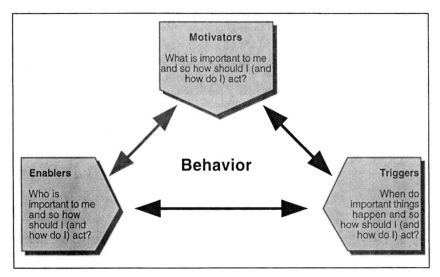

Exhibit 3-2. Guide for Designing Unwritten Rules Assessment Questions

as secondary to other business processes and, at worst, may be ignored. It is important to design systems that integrate the needs of quality with business requirements while minimizing features that exist solely to meet the needs of the Quality System. This concept is discussed further in *Regulations: Build a System or Add Layers?* (Ref. 3).

When trying to identify likely supporters or opponents, it is important to distinguish between two different categories of support or opposition:

- Issues that relate to the design of the existing systems or perceptions about *any* new system. Some people will be so content with the old system that they will simply not consider any change; others will be so frustrated they consider any change would be an improvement.
- Opposition or support to the idea of integration or to Quality Management in principle.

Approaches for these two categories must be handled differently. In addressing the first group, you should consider whether these opinions are going to influence others. If not, you probably should not spend too much time trying to influence them. Instead, work with others in their location who are ready to listen to you. Those who are in opposition in principle have probably had poor experience with comparable initiatives in the past. You should probe to find out what these experiences were and show them how this has, or will be addressed, in this project. If the opposition is particularly entrenched and you really must have those people's support, involve them in developing solutions.

Some managers are more likely to support your proposal than others:

- Managers who have recently been given ESH responsibilities are usually supporters of anything that will help them discharge their new responsibilities.
- Conversely, managers with established ESH roles may fear a reduced span of control.
- Managers with good experience of other Quality Management initiatives are likely candidates for active support.

In every location try to identify likely champions for your initiative, and be ready to provide them with the support needed to sway opponents.

◆**Hint**

Remember that support or opposition to your proposal, and the reasons behind these positions, will vary dramatically between individuals. Pay particular attention to the opinions of senior managers and other key "influencers." Make sure to address both the positive and negative issues in your plan.

3.3. Inventory and Assess All PSM, ESH, and Quality Management Programs and Elements

While you were securing support (Chapter 2) for the project, you prepared an initial list of PSM and ESH elements. This list now needs to be finalized and subjected to more analysis to identify overlaps between the various programs and elements. This is going to be the start point for redesigning some of the programs and elements. The initial list of programs and elements (Exhibit 2-7) can be included in the assessment tool kit as a prompt to the assessors or staff being interviewed. The team should also be looking for existing systems or processes that meet particular Quality Management requirements (Exhibit 1-1).

During this assessment phase, deliberately look for programs and elements. The most likely sources are Operating Manuals/Procedures, PSM, and ESH Manuals, and PSM and ESH specialists. In each case, your assessment team should seek out descriptions of the objectives of each

Exhibit 3-3: Sample Analysis of PSM and ESH Programs and Elements and ISO 9000

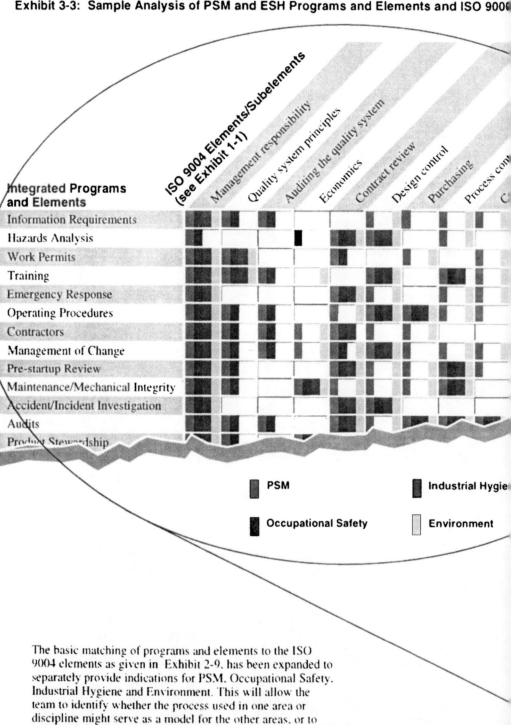

The basic matching of programs and elements to the ISO 9004 elements as given in Exhibit 2-9, has been expanded to separately provide indications for PSM, Occupational Safety, Industrial Hygiene and Environment. This will allow the team to identify whether the process used in one area or discipline might serve as a model for the other areas, or to see where three or four strong individual area programs must be integrated.

equirements for Xmple, Inc.

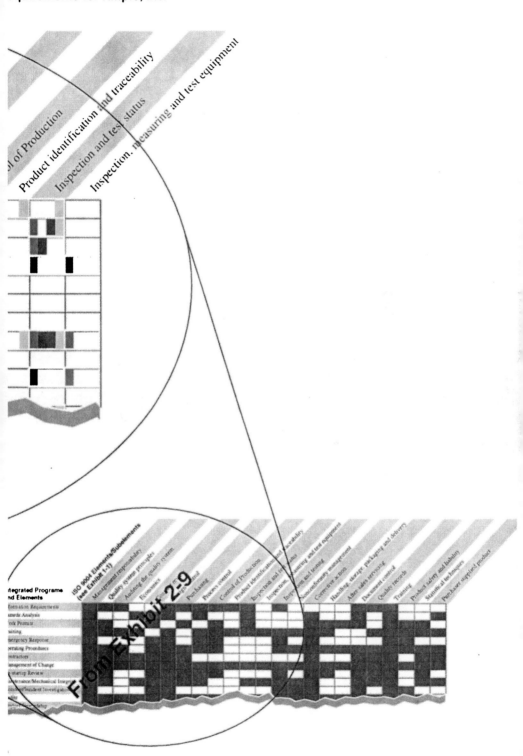

3.3. Inventory and Assess All PSM, ESH, and Quality Management Programs

program and element and get copies of any management systems or procedures. Your team should also try to form an opinion of the effectiveness of the program or element and record the basis for this opinion. The assessment should also consider whether a particular program or element could be adapted to meet the needs of other PSM or ESH activities. For example, could the spill prevention program in environmental management be extended to occupational safety, industrial hygiene and PSM? If so what changes, in broad terms, might be required?

◆Hint
Encourage your team of assessors to objectively review the effectiveness of existing management processes, programs, and elements. Make sure they gather as much "factual" evidence as possible, but don't ignore opinion and war stories given by interviewees. However, don't rely on the opinion of only one person; discreetly check with other interviewees for supporting opinions.

In the case of Cleanser Chemicals, Inc., the company formed a multifunctional team to assess commonalities across its existing Safety, Loss Prevention, Security, Medical, Industrial Hygiene and Environment functions. As a result of this assessment, many common requirements and potential work processes were identified, including:

Audits	Employee Training/Communicating
Capital Project Review	Government Relations
Community Outreach	Incident Investigation/Reporting
Contractor Program	Management of Change
Crisis Management	Management Systems
Emergency Planning	Process Design
Emergency Devices/Alarms	Program Documentation

These common requirements were then used to develop management standards, guidance documents, and new or revised procedures and work instructions for integrated ESH activities. In some cases, the integration went beyond ESH boundaries and also included engineering, maintenance, and technology centers.

The structure of the subsequent analysis of the programs and elements may depend on how you decide to build the new system:

- If each location will develop their own programs and elements, the analysis will be most useful if it is done location by location.
- If corporate or divisions are going to develop standard programs and elements, an overall assessment for the corporation or division will be needed.

In both cases the team should also expect to provide information on which locations have programs or elements that might be adapted, rather than built from scratch.

In either case, it is necessary to develop a more detailed chart describing what programs are in place. This can be based on Exhibit 2-9, which shows the relationship between the PSM and ESH programs and elements and ISO 9004 requirements. During the analysis, you need to consider each of the PSM and ESH systems separately asking a set of questions:

- Does this program or element exist today?
- What aspects of the current quality management system are already included in these systems?
- Which ISO 9000 requirements does it presently include?
- Could this activity be readily extended to cover other PSM or ESH activities?

The result of this analysis for Xmple, Inc. is illustrated in Exhibit 3-3 (see foldout), which provides a good visual impression of the strengths and weaknesses of the existing programs. This information will be used later (see Chapter 5) to decide how to start developing integrated systems.

3.4. Mapping the Management Processes

During this assessment phase, it will be necessary to develop an understanding of how the management processes for PSM and ESH flow through the organization. Your assessment should be designed to seek out this information. Two techniques for this analysis are discussed: process mapping and flow charting. Process mapping is used to understand how management processes flow through and across your organization. It helps identify inefficiencies, gaps in coverage or individual responsibilities. Flow charting is used to provide a clear picture of all the steps in the existing process and identify problems such as bottlenecks, repeated steps, missing

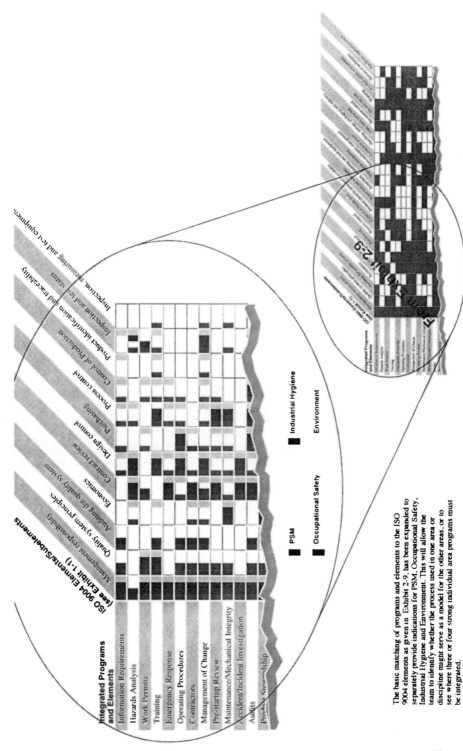

Exhibit 3-3. Sample Analysis of PSM and ESH Programs and Elements and ISO 9000 Requirements for Xmple, Inc.

steps and so on. Both of these techniques should also help identify existing quality management activities.

In some cases, you may find that process mapping is impossible or of little value. For example, it may quickly become evident that the gaps in coverage are so large that there will be little value to continuing with process mapping. In other cases, you may have process maps or flow charts already available to you. In such situations, the team will need to verify that the maps or charts are still current and complete.

For both process mapping and flow charting you can collect the information you need by asking each interviewee:

- Where do you obtain PSM or ESH related information or equipment?
- How do you use this information or equipment?
- Who do you pass PSM or ESH related information or equipment to?

3.4.1. Process Mapping

In process mapping you are interested in the way the process flows through the organization. Using the information gathered in your interviews you should be able to overlay the process onto an organization chart. This is illustrated in Exhibit 3-4. In Exhibit 3-4 several problems with the existing process are immediately obvious:

- The procedure is being developed without reference to the operators. Involving the operators is both good practice and a requirement of the OSHA PSM rule.
- The VP Operations–Olefins adds no value to the process, but is simply acting as a "post-box."
- The procedure is issued without proper approval and has few quality requirements included.
- There is an unofficial, but understandable, flow between corporate ESH and the local PSM coordinator.
- Contrary to Quality Management principles, the process is following organizational structures.

The process illustrated in Exhibit 3-4 is fairly typical of a traditional process that was designed without Quality Management processes and techniques built in.

3.4. Mapping the Management Processes

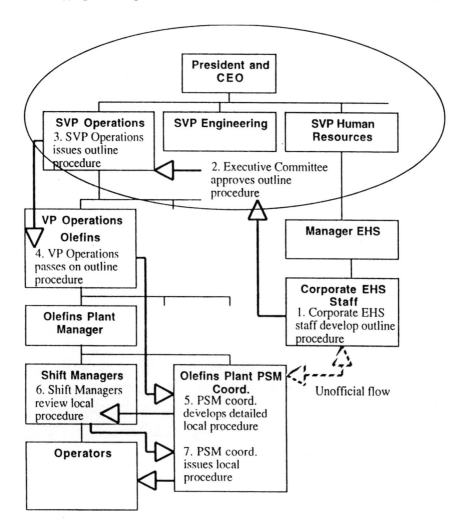

Exhibit 3-4. Sample of Management Process "Mapping" for Xmple, Inc. Developing and Issuing a PSM Procedure

---

◆Hint
In some organizations the management processes differ so wildly between departments, facilities, divisions, etc. that one process map will never make sense. If you find this in your organization it may not be worth continuing the mapping process. Instead, you should default to traditional flow diagrams.

3.4.2. Flow Charting

Flow charts may be developed through interviews or by assembling a group familiar with the existing processes. A facilitator helps the team identify all the process steps and builds up a representation of the overall process on a board using removable sticky notes. The resulting flow chart, Exhibit 3-5, has a more familiar look than that in Exhibit 3-4.

Examining the flow chart, some problems with the existing system become obvious:

- There are two steps where no processing takes place; both the SVP Operations and the VP Olefins are acting solely as intermediaries with no value added.
- The Olefins PSM Coordinator must work with five shift managers who are each required to review the procedure and return comments to him. This will almost certainly be a bottleneck in the process.
- As the process branches to send the outline procedure through the organization, there is a possibility that significantly different local procedures will be developed. For example, the VP Operations Olefins passes the outline procedure to five different olefins departments where each department will develop its own procedure. It is likely that a common olefins procedure could be developed, reducing the level of effort needed to develop five local procedures.

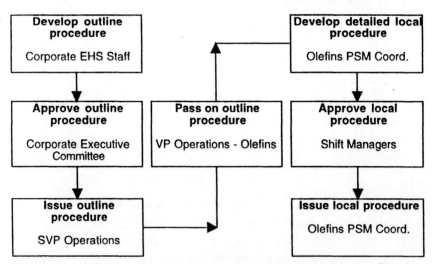

Exhibit 3-5. Sample Management Process Flow Chart for Xmple, Inc. Developing and Issuing a PSM Procedure

3.4.3. The Results of Process Mapping and Flow Charting

There are advantages and disadvantages to each method of developing management process flow diagrams.

- Process mapping will more clearly show duplications, nonproductive steps and gaps in coverage. But it takes longer to develop these diagrams and they are more difficult to read.
- Traditional process flow diagrams are clear, quickly prepared and easily reviewed for conformity with ISO 9000 requirements. However, it is not easy to identify some types of process breakdowns, such as gaps in coverage, processes that follow the organizational hierarchy and inappropriate allocation of tasks within the organization.

Since traditional process flow diagrams can be developed from the process maps, it is probably slightly better to use the mapping approach. If traditional flow diagrams are required they can be generated later to help identify redundancies, bottlenecks, etc.

◆Hint

Some process maps are very complex and may simply confuse the reader if they are included in the base plan. You may include them in an appendix, or simply keep them as reference documents in the project file. If you decide not to present them, it may be worth showing a simplified example so that your readers can understand the process you have used.

Process mapping also has the benefit of allowing Quality Management, PSM, and ESH responsibilities and accountabilities to be identified and analyzed. During the assessment, each of the Quality Management, PSM, and ESH responsibilities and accountabilities of everyone should be identified. All interviewees should be asked to describe their responsibilities and accountabilities. These can then be compared with their role in the existing management processes. Once again, if these are inconsistencies these will show up as mis-matches with the organization chart. For example, as shown in Exhibit 3-4, corporate ESH cannot be responsible for local procedures in Xmple, Inc. as they have no role in developing, no line management authority over and no oversight of, these procedures.

The assessment should also seek out systems for measuring performance. Remember that there is a need to measure both PSM and ESH performance, as well as the effectiveness of the processes in place. It is likely that at least some PSM and ESH performance measures will be in place, such as injury rates, number of incident investigations, regulatory infractions and insurance premiums. Your questions should probe for issues that managers track, but perhaps only informally, that could provide new measures of performance.

The efficiency of the existing systems is unlikely to be measured. Head counts of staff dedicated to PSM and ESH, and costs of PSM and ESH investments will be available. However, these provide no measure of efficiency and are probably inaccurate anyway. Other staff contribute part-time to PSM and ESH, and most projects will have benefits beyond PSM and ESH improvement. You should attempt to gather information on matters such as the time it takes to respond to new requirements and the annual costs of accidents, incidents, and noncompliances. These data will provide a baseline from which you can measure improvement as the integration project moves forward.

The team should also look for the processes which select priorities for investment and management attention. Possible drivers for these are:

- New regulations or industry standards
- Incident-, accident-, and noncompliance-related performance and costs
- Major incident/accident at competitor
- Rising insurance costs
- Risk assessment
- Pressure to reduce ESH and PSM costs

One useful by-product of improved PSM and ESH management can be a reduction in non–incident- or non–accident-related downtime. This arises because the improved systems provide the vehicle for improved mechanical integrity, better understanding of the design and operation and better statistical analysis of performance.

Finally, the assessment should identify the existing PSM and ESH resources, both manpower and equipment. Manpower estimates should include full-time staff and the full-time-equivalent of the part-time contributions made by other staff. You will need to probe for this information from part-time staff. Do not ask simply for the percentage of time they spend on PSM or ESH issues, as it is likely to be a wildly inaccurate estimate. Instead, ask about hours (not percentages of time) spent on PSM and ESH and other responsibilities such as plant operations, engineering, supervision, and planning. In this way, a more realistic estimate will be made.

Once the assessment is complete, coordinate your findings for use in integration (Chapter 5). The documentation will inevitably contain a great deal of data and should include supporting material such as survey results, audit findings, interview notes, flow charts, process maps, and evaluation of individual PSM and ESH systems. At this stage, you will need an executive summary of the findings to brief senior management and others. An example of part of such a briefing is provided at the end of this chapter in Attachment 3.1.

3.5. Redesigning the Management Systems

The assessment provides substantial information on the existing systems. Since it is not possible to redesign the entire integrated system simultaneously, you should start with the management processes, as these are the drivers.

The new processes should be designed in a series of quality workshops. The objective of the first workshop is to develop a generalized process that could be used as a starting point for detailed design of the final processes. For example, ISO 9000 provides a management process for implementing quality. The elements of this general process are the minimum requirements for the later processes. This workshop should include representatives of:

- PSM (both corporate and local)
- Occupational safety (both corporate and local)
- Industrial hygiene (both corporate and local)
- Environmental management (both corporate and local)
- Operations management
- Operations supervisor (for example, a shift foreman)
- Maintenance management
- Maintenance supervisor
- Quality Management system
- Others as required for your organization

Each quality workshop should be led by an experienced independent facilitator. Later workshops may involve a subset of this group.

At the initial workshop, someone from the integration project team should introduce the project and its objectives and then present the overall findings from the assessment.

♦Hint
When selecting the facilitator for the workshops, you should try to find someone who:

- Understands the quality management system you will be using
- Is independent of your team (a fallback position is a part-time member)
- Has experience in PSM or ESH management
- Is an experienced facilitator

The redesign should use Quality Management methodologies. The exact approaches will depend on the methodologies adopted by your company. However a possible sequence of activities would be:

1. *Define objective of the process.* The starting point for all Quality Management is deciding what the process is supposed to achieve. This may consist of primary and secondary requirements. It is likely to be a rather lengthy objective. For example, the primary requirement of a training program might be:
   To provide training in all aspects of PSM and ESH so that everyone in the company understands his or her responsibilities and has the knowledge needed to discharge these responsibilities successfully. Secondary training requirements would define the areas in which training is needed for every category of staff and employees.
2. *Use Fishbone diagrams* to identify all the process needs. This will draw heavily on the work to identify the PSM and ESH programs and elements and the Quality Management requirements. It will also identify any special expertise, information or equipment needs. Fishboning is described in several of the quality management references given in Chapter 1, and an example is provided in Exhibit 5-4.
3. *Use process flow modeling.* Using removable sticky notes, you can arrange the various requirements identified during the fishboning activity into a process flow model. This model needs to be carefully reviewed to make sure that it consists of the smallest number of steps necessary to deliver the required process.

Once a general process flow model has been developed, the workshop should conduct a series of "bench tests" to make sure it is robust. The group could be split into two or three smaller groups, each assigned to apply the process to one aspect of an integrated system, such as auditing or management of change. The groups should test that the specific requirements of PSM and ESH can be met by the general integrated process. When the general process passes these tests, it can be used as the basis of the detailed designs to follow.

3.6. Update the Implementation Plan

Once the assessment and general management process or requirements have been designed, the initial plan developed in Chapter 2 should be reassessed. Reexamine the manpower and schedule efforts in the light of what is now known about the existing systems and the final design. At this time, you should also be able to firm up your thoughts on your detailed implementation strategy by answering the following questions:

- *Can you develop corporate-wide systems or will you need different systems within each division or at each facility?*
  Factors that influence this include: the degree of decentralization to individual divisions or facilities and whether there are significant differences in organizational structure and business management processes between divisions and facilities.

- *Can you develop generic programs and elements that can be easily adapted to meet local needs, or will each location need to develop their own—perhaps fitting into a corporate framework?*
  Corporate-wide programs and elements can generally be maintained more easily with fewer resources. However, if there are significant differences in the risks, manufacturing processes, or overall management systems (non-PSM or ESH) between divisions or facilities, local programs and elements will most likely be required.

- *Which processes, programs, and elements should have highest priority?*
  In deciding where to set priorities, first pay attention to compliance with regulatory and internal requirements. Areas where risks are highest and areas where you already have good systems on which to model integrated systems should be considered next. Exhibit 3-3 will be very useful in identifying both gaps and areas of strength.

- *Which facilities should have highest priority?*
  Facilities with gaps in compliance or the highest risks should have highest priority. Other considerations might be poor existing performance or a recent deterioration in performance.

- *Are there obvious candidates for pilot studies?*
  In selecting a candidate for a pilot study, don't choose either the toughest or easiest candidate. In one case you are setting yourself up for failure, and in the other you may not adequately test your approach. Try to select a candidate where local management is supportive of your project, and where you expect to gain some obvious benefits.

When this reassessment is completed, report back to senior management and your other sponsors. You should let them know of any significant changes in direction, cost or schedule you have identified. If these changes are going to increase cost or lengthen schedule you should present alternative strategies that could reduce these. These might include:

- A reduced scope, including only a few divisions or facilities in the first wave.
- A longer schedule to reduce the resources required.
- Additional resources to shorten the schedule.

The team should also summarize the problems you anticipate with the project (from the unwritten rules analysis) and show how you propose to address these.

Explicitly ask for their support and the resources needed to move forward with the project. You should inform everyone of progress by widely communicating a summary of the material you presented to senior management and the outcome of the meeting. With this completed you are ready to move to the next stage of the project: developing a detailed plan.

References

1. Center for Chemical Process Safety of the American Institute of Chemical Engineers, *Guidelines for Implementing Process Safety Management Systems*, 1993.
2. Peter Scott-Morgan, *The Unwritten Rules of the Game*, McGraw-Hill, Inc., 1994.
3. A. M. Dowell, "Regulations: Build a System or Add Layers?," presented at the AIChE Spring National Meeting, Houston, Texas, March 21, 1995.

Attachment 3.1 Selected Slides from Executive Summary of the Assessment of Existing Systems for Xmple, Inc.

Executive Summary   Contents

This **Executive Summary** provides an overview of the findings of the assessment of the existing PSM and ESH systems

	Slides
• Major findings - summary	1 - 4
• Scope of assessment	5
• Approach	6 - 8
• Findings	9 - 15
• Next Steps	16

Findings - Summary

Integration will significantly improve overall management efficiency of PSM and ESH

- There are over 150 separate programs and elements in the existing PSM and ESH systems

- 70 percent of programs and elements are present in more than one aspect of PSM and ESH management

- There are 75 staff dedicated to PSM and ESH management
  - Part-time contributions are equivalent to another 120 full-time staff

- Management processes tend to follow the organizational structure
  - This slows the processes and increases the chance of breakdown

- PSM is the most complete system and already includes many elements of quallity management

- With the exception of the polymers division, there are four separate systems established for PSM and ESH management

1

Scope of Assessment

The assessment was in sufficient detail to provide an accurate indication of the current status of PSM and ESH management

- The Olefins and Additives divisions were subjected to a detailed assessment
  - All other divisions were subject to spot checks
- The assessment focused on identifying management processes
- The existing corporate audit was also used to gather information on the existing programs and elements

5

Approach   Overview

The assessment was conducted by using a preliminary survey followed by an audit

- A survey form was sent to every manager and supervisor. The survey sought information on:
  - What responsibilities each individual has for PSM and ESH
  - What quality management concepts were already in place
  - What programs and elements existed in their place of work
- Based on the survey we selected individuals at all levels for interviews as part of the audit process
  - During the audit we sought written and factual evidence on existing performance
  - We also determined the exact role of each individual in the management processes
- Following the audit we mapped the processes onto organization charts and prepared flowcharts for each process

6

Attachment 3.1

Findings Programs and Elements

There are over 150 separate programs and elements in the existing PSM and ESH systems

- The following counts of programs and elements were made
  PSM 15
  Occupational Safety 35
  Industrial Hygiene 40
  Environmental 60

- There are considerable inconsistencies in the way programs and elements are defined
  - Many programs and elements were developed locally, artificially increasing the total count
  - An initial attempt to derive common groupings indicates a total of 30 integrated programs and elements will be required

- For 28 of the 30 integrated programs and elements good models exist which can be adapted for integration
  - The remaining two programs (spill prevention and radiation monitoring) will need to be developed from scratch

9

Next Steps

We recommend that the implementation project go ahead with a target completion date 24 months from now

- Board approval required   Next meeting

- Establish corporate and divisional project teams   Month 1

- Select and engage consultant to assist in project   Month 2

- Conduct pilot project   Months 3-8

- Review pilot project   Month 9

- Implement in divisions   Months 10-24

16

4

Develop a Plan

4.1. The Need for Developing a Plan

An effective implementation plan for any new management system must consider a number of different issues:

- The plan should build on existing systems, where possible, since you are more likely to be successful if you can use proven systems. However, you should not adopt an existing system, regardless of how successful it is, if it cannot help with integration of PSM and ESH.
- The implementation strategy should be congruent with the objective of the project and the company organization and culture. The plan should be presented in a clear and succinct manner so that all those involved can follow it and progress against the plan can be monitored.

This chapter addresses all these issues. As with several of the earlier chapters, the overall approach to developing a plan should be modeled on that described in Chapter 5 of the CCPS publication *Guidelines for Implementing Process Safety Management Systems*. The approach recommended there is summarized in the following paragraphs.

Overall project approach

- While the appearance of the written plan can vary, and will most probably depend on your company's internal practices, the content should focus on:

    —*Statement of goals.* A simple statement indicating what systems will be included in the integration and what benefits will be derived.

    —*What is in place.* This should be a summary of the findings of the assessment.

    —*What is needed to develop integrated management processes and programs and elements.* This should be a summary of the integrated systems and programs and elements that will be developed.

    —*Tasks required.* A summary of the tasks that will be undertaken to achieve the objective, and the roles and responsibilities for each task.

    —*Work products or "deliverables".* A vision of what the integrated system will look like.

    —*Resources required.* Manpower and other resource requirements by task.

    —*Time required.* Usually a Gant chart showing the duration and inter-relationship between tasks. An alternative approach is a resource histogram showing trade-offs between staffing levels and the calendar time required.

    —*Expected results.* As far as possible a numerical description of the changes and improvements the integration will bring.

    Attachment 4.1 outlines two different sample plans.

- Review the preliminary plan that you presented to management. Compare the broad themes and projections you presented with the perspective you now have. If anything has changed dramatically, bring these to the attention of management sooner rather than later.

- There is no "correct" way to implement the integration project. You may choose a company-wide, facility-by-facility (or division-by-division) or hybrid approach. Factors that will influence your choice include:

    —any differences in the existing status of PSM and ESH between divisions, facilities or departments

    —how much PSM and ESH requirements differ between divisions, facilities or departments

    —whether company management is highly centralized or very localized

- *Company-wide approach.* A company-wide approach has several key benefits:

    —ensuring consistent systems, elements and programs throughout the company

    —minimizing resource requirements for development

    —the likelihood that the ongoing system can be supported by a relatively small, centralized group of experts

4.1. The Need for Developing a Plan

Even if you do not adopt a company-wide approach, you are likely to benefit from some central coordination to help share experiences and knowledge between locations. A company-wide approach is more likely to be adopted if:

—the current status of PSM and ESH management programs is relatively immature or weak throughout the organization. If this is the case it probably means there is not sufficient knowledge and expertise for local implementation.

—PSM and ESH issues are similar at each location. This is most likely to be the case if the manufacturing processes and materials are similar.

- *Facility or division-specific approach.* A facility or division-specific approach can provide very rapid integration and allow the systems, programs and elements to be adapted to meet local needs. However, the differences between locations will limit the value of sharing experience and knowledge. You are most likely to adopt a facility-specific approach if there are *significant* differences in:

—the current status of PSM and ESH management/programs
—the potential hazards at each location
—local regulatory requirements
—manufacturing processes
—process technology and equipment
—local management styles

Facility-specific approaches tend not to succeed where the overall status of PSM and ESH needs improvement or is immature in its development. Local staff do not have the necessary knowledge or experience to develop and implement the integration project.

- *Hybrid approach.* A hybrid approach addresses some matters centrally and some locally. For example, programs and elements may be developed locally due to significant differences in regulation and manufacturing processes, but the management systems (the framework of the integration) are developed centrally in order to establish uniform standards of performance. Management of hybrid approaches is particularly difficult and requires careful planning and well developed channels of communication.

Setting priorities

- Possible criteria for setting project priorities include:

—extent of compliance with regulations and corporate policy
—extent of potential hazards
—status of existing systems, programs and elements

This topic is also addressed in Chapter 3, Section 3.6.

- In setting priorities for action, the team must also consider available resources. You may identify an action step that produces the single largest improvement in performance, and then discover that it needs all of your resources. Alternatively, you may be able to spread your efforts among a number of lower priority actions that together have a greater impact on overall performance. For example, conducting combined PSM and ESH hazards assessments immediately might have the greatest immediate impact, but it would use all the available resources.

- Remember as you set priorities, that you and your team must consider the expected benefit to the company, division, or facility *as a whole*. Obviously, any specific immediate risk you identify in the course of your assessment must be dealt with quickly and responsibly; however, be careful not to let anomalous findings skew your perspective on broader priorities. For example, the absence of a management of change process is a significant gap. However, if you have very few changes it might have a lower priority than developing a comprehensive operator training program.

- When deciding which facilities have priority, consider the inherent risk (which depends on the type and quantity of hazardous materials and the conditions under which they are processed and stored) and the extent to which these risks are controlled.

- Defining priorities helps determine the scope of the plan. Remember that you may not be able to achieve total integration as a first step. Total integration may be achieved over an extended period and with a series of individual projects. You must design your plan so that subsequent integration projects can dovetail with yours.

Developing the plan

- A simple project management software package can be an invaluable aid to developing a good plan. These programs greatly reduce the effort required to modify plans and automatically make adjustments when variables change. Many good programs exist for both Macintosh and IBM PC compatible computers; ask someone in your company's planning or finance department for a recommendation.

- Having identified your priorities for action you need to develop a list of tasks that need to be undertaken, such as developing management processes, selecting pilot projects and communicating interim results. This can be done by developing a list of tasks yourself and having them reviewed by others, or you might assemble a group of engineers and managers and brainstorm a list of tasks.

- As you develop your network of tasks, you will quickly realize that not only are tasks within one priority area interrelated, but there are also relationships with other priority areas. These relationships will determine the order in which some tasks are completed and the resources you need to apply to complete them by a given date. For example, the team will need to develop programs and elements before they develop the contents of a training program, yet the training program itself may be a vital part of the process for developing individual responsibilities for that content.
- The team will need to staff tasks by matching skills with task requirements and will need to consider supervision needs for the task teams.
- The resource plan will be developed from the combination of the skills requirements and the task network. You may find a mismatch between your staffing needs and staff availability. When this happens you should consider:

  —do alternative resources exist?
  —is there another way to accomplish this task?
  —would a schedule adjustment free up needed resources?
  —what priority has been assigned to this task?
  —should we rethink that priority?

  If none of these provide a solution you should consider:

  —reassigning resources from another, lower priority task
  —hiring additional project staff
  —retaining consultants
- Communicate about your plan. You may need to obtain formal authorization to continue with the project from top management. You should also remember that other managers may influence the decisions of top management. You should consider informally reviewing your plan with selected members of this group. Once your plan is approved you should communicate widely on the general points of your plan. The assessment of existing systems may have provided some interesting material that could be used for internal articles.

4.2. Adjust the Preliminary Plan

Having completed the assessment of the existing systems, you will now be in a position to estimate the level of effort required to design and implement new systems. You have now identified where systems, programs and elements exist and their potential value as a basis for integration. In other areas, you will have nothing to work with and will need to design completely new systems. Clearly the level of effort required will vary dramatically depending on a number of factors, including:

- whether a particular aspect of integration will be based on an existing system or a completely new design will be required. For example, it is unlikely that many companies already have measures of management efficiency for PSM and ESH in place, but they all measure end-of-pipe performance.
- for an existing system, program or element that is to be adapted: the degree to which it already conforms to your quality management system.
- the completeness of an existing system, program or element. Does it include features that will easily be adapted to meet the needs of other PSM or ESH areas, or will additional features be needed? For example, PSM management of change requirements that are very comprehensive would be easily extended to cover the other ESH areas.

Exhibit 4-1 provides some indications of the level of effort that might be required to develop integrated systems, programs and elements.

The level of effort estimates given in Exhibit 4-1 are for time used by the project team. Design time is an estimate of the number of person days that will be needed to develop a fully defined system program/element or management process starting with the data collected during the assessment phase.

Exhibit 4-1. Level of Effort Guidelines

	Development of Integrated Program or Element (from Exhibit 3-3)	Development of Associated Management Process
Good Model Exists to Use for Integrated Design	5–10 days for design 5–20 days for buy-in	5–15 days for design 10–20 days for buy-in
Model Exists, but Has Deficiencies to Resolve	10–15 days for design 5–20 days for buy-in	15–20 days for design 10–20 days for buy-in
No Model Exists and Must Develop One from Scratch	10–30 days for design 5–20 days for buy-in	15–25 days for design 10–30 days for buy-in

4.2. Adjust the Preliminary Plan

- A good model would be something that is fully developed in at least one area of PSM or ESH, and can readily be adapted to the other areas with few additional requirements.
- A model would be something that exists in one area of PSM or ESH, but will require significant upgrading before it could be adapted for use in other areas.
- No model is considered to exist where the work required to adapt an existing model exceeds that needed to develop a new one.

The time for buy-in is for presentation and negotiation with individual facilities, divisions, etc. to make sure the component will meet their needs. You should not underestimate this requirement as it will contain elements of training, explanation and negotiation.

The low end of the time estimates should be used where the component will apply across only a small number of different divisions, facilities and units—less than three, say—and where the assessment has shown a high degree of acceptance for integration. The high-end estimates should be used where a large number of divisions or other groups will be included, where there are differing hazards or if there is reluctance to accept integration.

◆Hint
Take advantage of corporate experience with other Quality Management projects when preparing your cost estimate. This will provide actual costs to use as a basis for estimating the costs for developing management systems and programs.

During the assessment of existing systems, Chapter 3, you will have developed a detailed understanding of each of the components of the existing management systems. During the planning phase you want to minimize the number of completely new components you develop. By building on existing material, you maximize the chances of developing a system that will work within your organization. It is now necessary to choose your building blocks for the new systems. Considerations include:

- For programs and elements the number of areas of PSM and ESH that include a requirement:
  —If only one area requires a program or element, there is no need for development.
  —Where two or more areas include a requirement, you will need to

select an existing program to build from. Generally you should select the area that has the most comprehensive program or element as your building block. It is generally easier to simplify or eliminate existing requirements than to build new ones.
- For management processes, you should identify the individual management activities that exist and are needed in the integrated system. You should evaluate each of the existing management processes looking for similarities and differences, as well as strengths and weaknesses. Based on this, you can select a set of management activities as building blocks for the integrated system.
- It is important to make sure that all the needs of each area of PSM and ESH are addressed in the integrated system. You should conduct a paper exercise where you step through the existing components and the new components in parallel, making sure you have captured everything—or at least everything you need (there may be extraneous activities in the existing components).

One company found that it had some common improvement opportunities in the area of incident investigation. The occupational safety group had already discovered that it had site-to-site inconsistencies in conducting accident/incident investigations when the process safety group determined that they needed a better accident /incident investigation system under OSHA 1910.119. The two groups were able to jointly decide on a model and the associated training requirements, creating an efficiency in both the planning and execution stages.

Inevitably some responsibilities for PSM and ESH will need to shift to different managers and staff. The changeover is a potential problem both because there is the possibility of something falling through the cracks and because of resentment over losing or gaining responsibilities. The issue of resentment should have been identified during the "unwritten rules" assessment and the steps to address these identified. Preventing anything falling through the cracks comes down to careful planning:

- Remember that it is almost impossible to transfer responsibilities instantly. However good the training, those picking up new responsibilities need a handover period.
- Consider the following strategies for handover:
    —Have the newly responsible staff work alongside the existing staff under the old system with the existing staff retaining their responsibilities.

—Have the existing staff work alongside the new staff under the new system with the new staff taking over responsibility.
—A combination of both of the above.
—Design an interim management process that includes both the existing and new staff.

In choosing your strategy for handover, it is necessary to consider your company's experience with other new systems. However, bear in mind that some handover issues will be very local in nature and one solution may not work in all instances.

4.3. Implementation Strategy

The implementation strategy will need to be consistent with your company practices at both the corporate and local level. Consider the following:

- *Project approach.* Your company may have established project management practices. They may be described in a Project Management manual. Familiarize yourself with these practices, perhaps arrange to attend an internal or external training course, if that is appropriate.

    If you do not have corporate practices, you will certainly have corporate experience which should be reviewed. Experience with quality projects will be the most relevant. In some instances, formal postproject reviews will have been conducted and these reports will provide valuable insight for the development of your project. In the absence of such reports, you should arrange to spend some time reviewing previous projects with managers and staff who were involved. Two samples of a table of contents for a project description/plan are given in Attachment 4.1.

    Focus your attention on project organization, project planning process and project monitoring and reporting systems. The team should specifically be asking what worked well, what needed to be fixed and with hindsight what would have been done differently. As you review existing experience, you should remember that your project objectives are different; does this change your assessment of past projects?

- *Pilot project.* An appropriate pilot project can be extremely beneficial to the overall success of the integration project (see Chapter 6 for more information). In particular:

— A pilot project allows you to debug your design and project management processes.
—It can provide a demonstration of the benefits that will be achieved.
—It can demonstrate the costs of the integration project.
—It can create advocates of the integrated approach.

A pilot project generally ensures the complete project is implemented smoothly and a clear understanding of the value of the project is established. Of course, a failure at the pilot stage can lead to cancellation of the overall project. If the pilot is not successful, it is important to analyze the reasons for failure so that any decision to continue or abandon the integration is made rationally.

Selecting the scope and location for the pilot project is important. The pilot must provide a realistic test for integration, without being unrealistically ambitious. The ideal subject for a pilot will have staff who are motivated to make integration work, yet have obvious inadequacies in their existing PSM and ESH management.

The pilot study could attempt to develop a totally integrated management system for all aspects of PSM and ESH management. This may be possible if a small independent business unit can be identified. However, it is more likely that you will select one or two aspects of PSM and ESH management at one location for the trial. You should choose management systems that are self-standing such as training or measurement. The programs or elements you choose for the pilot should also be as independent as possible, for instance, spill prevention or selected permit-to-work procedures. You should avoid systems that are highly integrated with other activities, for example, management of change.

- *Full or partial implementation.* For an integration project, partial or full implementation is relevant to which locations, divisions and facilities will be covered. Furthermore, if your company or a particular division is highly centralized, partial implementation may be out of the question. The central resources will not be aligned to provide support to both integrated and unintegrated activities except by duplication of resources.

  If the potential hazards or manufacturing processes vary between entities within your company, partial implementation may be a real consideration. Low hazard activities will not enjoy the same

4.3. Implementation Strategy

level of benefit from integration that high hazard activities will. This relates directly to the level of effort needed to manage PSM and ESH under the existing arrangements. Review the benefits each entity will gain against the costs of integration, as you may find some entities within the company for which integration makes no sense.

You may also decide to implement integration step-wise, where each step integrates a whole division. This approach can be successful if PSM and ESH is managed at divisional level. This has the advantage of requiring fewer resources, but will lengthen the time required to achieve all the benefits.

- *Introducing the new systems.* Redesigned elements and programs can be introduced under the existing management systems. This can result in the same program or element being managed under several different areas of PSM and ESH. Integrated management processes cannot be introduced ahead of the programs and elements. The integrated management processes have been designed specifically to manage integrated programs and elements across all areas of PSM and ESH.

  Within the management processes some are self-standing, others are closely related with one another. You have used your quality management system as the basis for the new processes. Review each requirement of your quality management system to identify self-standing and grouped requirements. Exhibit 4-2 shows possible self-standing and grouped requirements for ISO 9000.

  Within ISO 9000 there are two requirements that are general to all other requirements and must be designed into each requirement. There are a small number of truly self-standing requirements that could be implemented at any time. The majority of the ISO 9000 requirements need to be implemented in a specific sequence and within this sequence there are some requirements that must be introduced as groups. The sequencing shown in Exhibit 4-2 is not sacrosanct; some variations from this are quite likely, particularly where you already have some aspects of Quality Management in place. It is important to conduct a review of Quality Management system requirements against the assessment of existing systems.

- *Quality tools.* During the integration project you will make use of a number of different Quality Management tools. Most of these will be used during the design phases of the project. Tools you are most likely to use are:

Exhibit 4-2. Groupings of ISO 9000 Requirements for Implementation

---

Basic requirements applicable to all management systems
    Management responsibility
    Quality system principles

Self-standing—can be implemented at any time
    Contract review
    Inspection, measuring and test equipment
    Handling, storage, packaging and delivery
    After-sales servicing
    Quality records
    Training
    Product safety and liability

Grouped and related requirements—must be implemented in groups and in sequence specified below

    Design control
    Purchasing
    Process control
    Control of production
    Purchaser supplied equipment

    Material control and traceability

    Inspection and test status
    Inspection and testing

    Nonconformity
    Corrective action

    Quality documentation and records

    Use of statistical methods

    Economics

    Auditing

    —*Fishbone diagrams* to help identify detailed requirements. Most likely to be used during development work within the project team and during consultation within your company.
    —*Flow charting* is used to help understand existing systems and to design new systems.
    —*Process mapping* is used to understand how management proc-

esses flow through and across your organization. It helps develop efficient processes that reach all the relevant parts of your organization and also helps define individual responsibilities.

—*Benchmarking* is used to identify best practices for consideration in your integration project. Benchmarking can be applied to internal processes (comparing management processes or programs and elements at different locations or areas of PSM and ESH) or to external organizations.

◆Hint

In developing your implementation strategy, you should review experience with other projects, particularly Quality Management projects.

4.4. Update Benefits and Costs

Once you have developed a detailed plan, you should once again revisit the expected benefits and costs. If anything has changed significantly, you will need to bring this to the attention of senior management as early as possible.

Following the assessment of existing systems, you will have been able to update your projection of the potential benefits. For example:

- Improved compliance with regulations and internal policy
- Faster response to changes in regulation, products, organization and manufacturing methods
- Fewer steps in management processes
- Fewer resources required to manage PSM and ESH

Once the detailed project plan has been developed, your estimates of costs and schedule in a number of areas will have firmed up:

- Resources required for design and installation of integrated system
- Resources left in place to manage the on-going system
- Overall project schedule and expected timing of benefits

It is common for there to be some significant changes between your earlier estimates and those made after the detailed plan has been developed. For example:

- You might have projected that some benefits would be available across the company within 12 months of project initiation. However, you have now concluded that the project will be implemented step-wise one division at a time. The result of this is that although the first benefits will still be seen within one year, the total benefit will not be seen until the third year.
- Your project resource requirements are lower. Following the assessment of existing systems you found that for most of the programs and elements good models existed that can be readily adapted for general use.

These are examples of changes that may have occurred. When presenting any changes in your cost/benefit projection you should, as far as possible, present numerical data rather than judgments.

4.5. Recast the Plan

Once all the detailed inputs for the plan are developed, the team can update and elaborate on the formal plan presentation. The introduction to this chapter provided a complete list of the required content of the plan and some thoughts on how to develop it in more detail. One aspect not covered is integration into business plans for each business unit.

Business unit managers are usually more receptive to costs that they can plan for ahead of time. If their resource requirements or cost base is going to change, it is easiest to include this during the normal annual planning cycle. Make sure you understand the planning cycle and try to time your project plan and cost estimates so that they can be included in the annual planning cycle. Remember that you will be competing with other projects for inclusion and your project must demonstrate sufficient benefits if it is to be included.

Usually, there will be a department or individual with a central role in the planning process. You should spend time with these staff developing an understanding both of the planning process and the evaluation criteria they use for including expenditures in their budgets. If your project has top level backing, some costs may be absorbed by corporate; this is most likely to be the case if costs are incurred well ahead of any benefits.

If the timing of your project is such that some costs will be incurred ahead of the annual planning cycle, you may find yourself competing for very limited discretionary funding. In these cases, you may want to try to structure your plan so that these initial costs are kept as low as possible. You can then have the remaining costs included in the normal planning cycle.

◆Hint
Keep communicating! At this stage it may be impossible to do too much to keep key managers informed.

Once the plan is complete, you will almost certainly need to submit it for final approval. Your company probably has well established approval processes. Make sure to understand these and adequately brief anyone who may present your project to project review committees or board meetings on your behalf. When preparing your plan for approval, it is important to briefly reiterate the history of the project. Although you have thoroughly prepared senior management through your various communication and consultation exercises, they will not recall all of the relevant background.

Also remember to be specific about what approvals you are seeking. If you are asking only for partial funding make that clear, but also make it clear that this is only one step on the road and additional funding will be required later. If you are asking for full funding where the costs may be large, point out that no additional expenditures are expected. You should also be sure to address the risks associated with the project: How certain are the benefits and costs? What might slow down the project and the flow of benefits? Does the project depend on other parallel projects? and so on.

Throughout the approval process, work with your senior champions to make sure they have the information they need to keep the project moving. Listen for the sounds of opposition and move to defuse any issues that may obstruct the project. Don't assume no news is good news. Bad news is difficult to deliver; if you expected to hear something on a particular day and didn't, you should call and find out what is happening!

Reference

1. Center for Chemical Process Safety of the American Institute of Chemical Engineers, *Guidelines for Implementing Process Safety Management Systems*, 1993.

Attachment 4.1. Sample Plans/Project Descriptions

The following is an outline of one company's plan.

1. **Organization and Responsibilities**
    1.1 Project Description
       1.1.1 Outline of Project Objectives
       1.1.2 Summary of Costs and Benefits
       1.1.3 Outline of Implementation Strategy
       1.1.4 Project Approvals
    1.2 Project Organization
       1.2.1 Corporate Teams
       1.2.2 Divisional Teams
       1.2.3 Funding Arrangements
    1.3 Project Responsibilities
       1.3.1 Executive Oversight
       1.3.2 Project Management
       1.3.3 Project Change Procedure
2. **Administration**
    2.1 Communications
       2.1.1 Internal Project Communications
       2.1.2 Internal Company Communications
       2.1.4 External Communications
       2.1.5 Overall Communications Schedule
    2.2 Documentation
       2.2.1 Required Documents
       2.2.2 Document Approval
       2.2.3 Document Distribution
    2.3 Progress Reporting
       2.3.1 Standard Progress Reporting
       2.3.2 Exceptions Reporting
3. **Project Controls**
    3.1 Work Breakdown Structure
       3.1.1 Central Project Tasks
       3.1.2 Divisional Project Tasks
    3.2 Schedule
       3.2.1 Overall Project
       3.2.2 Pilot Project
       3.2.3 Divisional Projects

3.3 Resources Needed
    3.3.1 Central Project Team
    3.3.2 Divisional Project Team
3.4 Resource Availability
    3.4.1 Corporate Resources
    3.4.2 Divisional Resources
    3.4.3 External Resources
    3.4.4 Resourcing Procedures and Approvals

4. Quality Plan
    4.1 Quality Management Objectives
        4.1.1 ISO 9004 Requirements
    4.2 Project Quality Measurement

5. Installation and Verification
    5.1 Pilot Testing
        5.1.1 PSM and ESH Measures of Performance
        5.1.2 Process Efficiency Measures
    5.2 Training
        5.2.1 Project Team
        5.2.2 Others
    5.3 Implementation and Interim Arrangements
        5.3.1 Introduction of New Programs, Elements and Management Processes
        5.3.2 Interim Arrangements During Changeover
        5.3.3 Phase-out of Old Arrangements
    5.4 Verification
        5.4.1 Pilot Project
        5.4.2 Divisional Projects

6. Financial Controls

Another sample is provided below.

I. Introduction and Background
    A. Motivation for Change
    B. Departments and Functions Covered
        1. Corporate ESH Functions
        2. Facility ESH Functions
        3. Other Functions
    C. Anticipated Benefits and Costs
        1. Funding for Integration Effort

2. Short-Term Benefits and Costs
   3. Long-Term Benefits and Costs
   D. Integration Team
II. Description of Current Status
   A. Number of Programs and Elements Today
   B. Apparent Inefficiencies and Redundancies
   C. Existing Systems to be Incorporated in New Design
III. Integrated Design
   A. Integrated Systems to be Developed
   B. Integrated Programs and Elements to be Developed
   C. Description of Expected Results
IV. Work Tasks
   A. Task 1
      1. Description of Task 1
      2. Roles and Responsibilities
      3. Work Products/Deliverables
      4. Resources Required
      5. Schedule for Task 1
   B. Task 2
      1. Description of Task 2
      2. Roles and Responsibilities
      3. Work Products/Deliverables
      4. Resources Required
      5. Schedule for Task 2
        .
        .
        .
   N. Task N
      1. Description of Task N
      2. Roles and Responsibilities
      3. Work Products/Deliverables
      4. Resources Required
      5. Schedule for Task N
   O. Overall Schedule
      1. Gant Chart
      2. Key Dependencies and Inter-relationships
V. Progress Reports and Communication

5

Integration Framework

5.1. The Need for Developing an Integration Framework

The framework for integration provides the skeleton on which the complete system will be built. The framework defines the overall structure of the integrated systems, the way they will be built and which tools will be used to build them. Correctly designed, the framework will ensure that the integrated PSM and ESH systems will match other management systems in your organization and meet the requirements of ISO 9000 or your corporate Quality Management system. The framework will also make sure that the new systems will be developed and implemented in a rational sequence. This chapter describes some of the design issues and Quality Management tools that are available. Suggestions about how to structure the integration framework are also provided.

This chapter first discusses how to set about prioritizing your integration efforts, then how to develop integrated systems and build the concept of continuous improvement into your systems. A section is also provided on the various tools that might be used in the process. Lastly, there is a section on how to approach integration if you are dealing with informal existing systems. This section will not apply to all readers.

During the analysis of the existing systems, the team will have learned a great deal about how PSM and ESH are currently managed. This analysis should allow you to identify the differences between and inconsistencies within the processes in existing programs. You should review these differences and inconsistencies and decide whether they are substantial or trivial. Where substantial differences are uncovered, you will need to make sure that you understand why these differences exist and take account of these in designing the integrated system. Where differences are justified, you may need to design these into your integrated system. In centralized organizations you will need to develop central systems that allow for the appropriate degree of local customizing. For example, you might set policy and scope of coverage centrally and allow divisions or local management to develop or adopt appropriate procedures and standards.

5.2. Prioritization of Programs, Elements, and Processes for Installation

There are no hard and fast rules for deciding the order in which to install programs and elements. But you should develop and install the management processes first—without these the programs and elements would be working in a vacuum and would most likely be ignored (everything will carry on as before). For example, a change control process should be developed and installed before management of change procedures are established.

♦Hint
- Develop and install management processes before programs and elements, as this provides a management structure within which to manage the programs and elements. For example, you should establish management responsibilities before developing detailed management of change procedures (also see Exhibit 5-1 which provides guidance on priorities for installation).
- In deciding which order to develop and install the integrated systems, look for a balance of early success and enough challenge to demonstrate the benefits of integration.

ISO 9004 defines 23 Quality Management system elements (see Exhibit 1-1) which have been grouped for implementation (see Exhibit 4-2). The priority for implementation must now be decided. Exhibit 5-1 suggests priorities for the groups in Exhibit 4-2 and some considerations that will determine the priorities in a particular integration project. The actual priorities in any particular case will vary depending on local considerations. For example, *contract review* may be installed early if there is a great deal of project activity or frequent use of contract labor. Your version of Exhibit 3-3 may provide additional ideas and guidance.

5.2. Prioritization of Programs, Elements, and Processes for Installation

Exhibit 5-1. Possible Priority for Installation Using ISO 9004 (shown in decreasing order of priority)

ISO 9004 Grouping	Discussion
Management Responsibility Quality System Principles	These two requirements are fundamental to any quality-based management system and must be the first to be installed.
Design Control Purchasing Process Control Control of Production	This group of requirements ensures that equipment is correctly designed and operated. Other elements will ensure compliance with the output from this group.
Inspection and test status Inspection and testing Inspection and test equipment	In the chemical industry, verifying that all equipment is fit for operation is a critical requirement for continued safe operation. Inspection and test equipment must be available.
Auditing	While inspection verifies the condition of equipment, auditing verifies procedural and management system requirements. Installation now should focus on existing systems. Other audit elements can be added later.
Nonconformity Corrective Action	Once design and operational requirements have been set, and inspection and audit put into place, a system to correct deficiencies should be developed.
Training	Training programs can only be developed after training requirements are established. In reality, training will be a continuous requirement, but this represents a point at which some significant new requirements have been put in place.
Material control and traceability	This item could have a higher priority if raw material or product losses are high or represent a significant safety or environmental risk.
Quality records	A critical element in improving performance is establishing measurements of performance which can subsequently be analyzed to identify opportunities for improvement.
Use of statistical methods	Follows logically from establishing quality records.

Continued on page 94

Exhibit 5-1. *(continued)*

ISO 9004 Grouping	Discussion
Economics	No quality system makes sense unless it is contributing to improvement in economic performance. However, until the bulk of the quality system is in place it cannot be effective.
Contract review	This requirement will have a much higher priority if you use contract labor routinely or are likely to have project work at your facility. This review ensures that PSM, ESH, and quality have been adequately addressed in all contracts.
Quality documentation and records	This requirement should be developed during the whole of the installation process. You need to document your system. However, you must carefully consider the timing of publication. In some projects it may be appropriate to issue documentation from day one and routinely update it.
Handling, storage, packaging, and delivery After-sales servicing Product safety and liability Purchaser supplied equipment	These requirements may not apply to all chemical manufacturing facilities. Where they are required, they are most likely the partial responsibility of the marketing function and you should work with them to establish relevant priorities.

A cross-reference table for the requirements listed in Exhibits 4-2 and 5-1 and the ISO 9000 series is provided in Exhibit 1-1. Appendix A provides descriptions of each requirement of ISO 9004.

Setting priorities for developing and installing programs and elements should address the following considerations:

- *Likelihood of success and degree of difficulty.* It is important to gain some early success that will provide credibility to the integration project, and these successes must be selected and planned. However, if the team's target is too easy it may be dismissed as being unrepresentative. You should try to select a mixture of easy programs and elements and some offering a greater challenge for early installation. Working on challenging assignments early has the advantage of

5.2. Prioritization of Programs, Elements, and Processes for Installation

Working on challenging assignments early has the advantage of allowing you to identify and overcome problems early in the project, and to harness the initial burst of energy and attention that such projects typically receive.

For example, *hazards assessments* programs might be an early priority. Although a good Process Hazards Assessment program may exist within PSM, there may not be programs of comparable quality in occupational safety (job safety analysis) or environmental management. Developing an integrated program will provide early benefit to the occupational safety and environmental management. Using the existing PSM program as a basis will likely ensure success, yet there will be enough challenge from integrating other ESH needs for *hazards assessment* to be a credible test of integration.

- *Existing strength or weakness.* A strong program or element will generally be well understood by everyone. This provides an interesting challenge—introducing change may be difficult as there will be reluctance to change "something that is already working". However, the strong programs and elements are also those least likely to require significant change. You should assess the degree of change that may be needed and the likely resistance to such change. For example, if you have a strong environmental management training program, integrating the PSM and other ESH programs may be relatively easy using the existing environmental management framework.

   Where you identify particularly weak programs or elements, these will frequently be good choices as high priority items. A weak program or element is likely to be recognized as such by managers and they will enthusiastically support its development and installation. For example, there may be different audits conducted by different specialists or functions for PSM, environment, occupational safety and industrial hygiene. An integrated program offers managers fewer audits and will be less distracting for their staff.

- *Compliance or risk management priority.* The earlier assessment of existing systems may have uncovered regulatory requirements that are not being met or risks which are not controlled. Programs or elements that correct such oversights must be given the highest priority. For example, your PSM process safety information may not meet the needs of the regulation. If this is the case, it may make sense to develop a single information program covering all areas of ESH and PSM.

One company elected to integrate the management of change process first, as the integration exercise itself represented change, and they felt that working on any other single process would still leave the overall operation vulnerable during integration. This company felt that change represented the only vehicle for enhanced ESH performance, but that it was also a major threat to sustained performance, if not managed properly. By first integrating their variety of change control mechanisms into one robust but simple process, the likelihood of undermining prevention controls during subsequent integration was greatly diminished. Consolidation of follow-on processes became easier by virtue of being able to address proposed changes related to integration by a single method—thus serving as an instantaneous pilot test! Immediate results were visible in the form of higher quality and more rapid and less costly handling of current challenges, issues, and opportunities.

5.3. Developing Integrated Systems

Once the order in which you plan to develop management processes, programs, and elements has been established, you must assign responsibilities for the work. These will be shared between the project team and local managers and staff. Generally, small development teams will be established to work on particular processes, programs, and elements. Ideally each team should include the following:

- Quality Management experience and meeting facilitation skills
- PSM and ESH expertise
- Experience of relevant technical disciplines
- Experience of relevant operations activities
- Experience of relevant inspection or maintenance activities
- Management system development experience
- Representative of group who will "own" the final product—the owners are the group who will use and maintain the final product.

Different requirements may be filled by a single team member. One member of the team should be appointed as the team leader and the team should agree about the division of other responsibilities. The team leader should be given clear instructions on the scope of work for the team and the schedule within which they are working. This will most likely be achieved through the application of Quality Management analysis techniques. Staff can be involved in more than one team, but avoid having one person lead more than one team. Much of the work of the team will be conducted in team meetings during which the team will analyze and specify various requirements. Individual members of the team will be responsible for preparing various materials such as draft policies, procedures, and standards.

Once the process, program or element has been developed and thoroughly reviewed, installation can go ahead. Although there should be continuity between the development and installation teams, they are unlikely to be the same staff. Installation is best managed by those who will actually be using the process, program or element. Installation must include any interim arrangements that are needed during the changeover from existing to integrated systems. This will involve a significant level of effort and may also take a considerable length of time.

Successful implementation requires that the new system is reviewed by all those who will use it, or by representatives of these groups. Generally a small team will be appointed with overall responsibility for implementation; this team will include representatives from every major group that will use the new system. Once this team has agreed on the overall implementation strategy, each member of the team will work with one or more user groups to explain the new system and its implementation.

Exhibit 5-2 shows three sample project installation strategies. Each example shows one possible implementation strategy. Variations on these will almost certainly be required to meet local circumstances. Example 1 envisages shared responsibility for the project with local staff. Example 2 shows local staff taking the lead and Example 3 shows minimal involvement of local staff. Other combinations of responsibilities and the use of other resources are also possible. Constant in all the examples is the development of management processes ahead of programs and elements, and the provision of local training before installation starts.

5.4. Continuous Improvement

Building continuous improvement into management processes is a vital element of the integration program if the system is to withstand the test of time. Almost inevitably, regulations, manufacturing processes, internal organization and management processes change over time. Additionally our knowledge of environmental, health, and safety hazards continually improves and we add new techniques to our PSM and ESH toolbox. Finally, however good the initial design of the integrated system, opportunities to improve it will be found.

5. Integration Framework

Example 1

- Integration Team develops Integrated Management Processes
- Local staff training on new Management Processes
- Measurement of existing performance
- Integrated Programs and Elements developed by Integration Team and Local Staff
- Operator training
- Interim arrangements developed
- Interim arrangements installed
- New management processes installed
- New programs and elements installed
- Project review to assess pilot project

Example 2

- Local staff trained on Integration approach
- Integration Team and local staff develop Integrated Management Processes
- Local staff develop integrated programs and elements
- Measurement of existing performance
- Operator training
- Interim arrangements developed
- Interim arrangements installed
- New management processes installed
- New programs and elements installed
- Project review to assess pilot project

Example 3

- Integration Team develops Integrated Management Processes
- Integrated Programs and Elements developed by Integration Team
- Interim arrangements developed
- Local staff trained and proposed systems reviewed with them
- Operator training
- Measurement of existing performance
- Interim arrangements installed
- New management processes installed
- New programs and elements installed
- Project review to assess pilot project

Exhibit 5-2. Example project development and installation strategies

---

◆Hint

Try to avoid bolting on continuous improvement, rather make it an integral part of the systems you put in place. If it is bolted on, it is probably not essential for day-to-day management and could quickly be ignored. For example, you might decide that the overall system will be reviewed every two years to identify improvement opportunities; this can too easily be postponed, forgotten or abandoned. Instead, the system should be designed to identify weaknesses and improvement opportunities at any time. In other words, continuous improvement must be built-in.

5.4. Continuous Improvement

There are several components to continuous improvement, all of which should be included in the overall framework:

- *Process to identify and respond to changes.* The team should incorporate a process or processes to identify changes in regulation. Such monitoring will allow the organization to plan early and minimize the costs of compliance. For example, the current PSM regulations in the United States impose certain process hazard assessment (PHA) requirements, while proposed amendments to the Clean Air Act impose some different PHA requirements. A company that includes the new requirements in PHAs being conducted today may avoid the need to conduct new assessments in the future. You cannot directly influence regulation, but you should monitor proposed new regulations and their progress to implementation. They will also be in a better position to lobby for change, either directly or through an industry association.

    Most other influences for change are driven by the internal decisions of your management. If the company has adopted Quality Management principles throughout the organization, you will be given time to redesign your management systems. However, the team should anticipate the need for change even as you design new processes. Make sure you develop process flow diagrams and descriptions of each step. In this way as management responsibilities or manufacturing processes change, responsibilities for each process step can quickly be reassigned. It is prudent to repeat the process mapping onto any new organization to check that no one has been left uncovered by the new process.

- *Plan, do, check, act cycle.* The basic structure of a Quality Management process includes these four steps, which ensure that any deficiencies in the existing system will be identified and corrected. Every management process must have steps that identify what need to be done (plan), an implementation phase (do), a measurement and/or review activity (check) to make sure the system is working as expected and an action step (act) to rectify any problems. While not all these steps are active at any time, the overall management process should ensure that all these activities take place on a regular, periodic basis. The developed framework should require each management process to include these four activities.

- *Measurement of performance.* Quality Management requires that measures of performance be established for every activity. These measures include end-of-pipe measurement, such as amounts of material released into the environment or injury rates, and in-process measures of how efficiently you are managing, such as time to review safety improvement proposals or total resources expended on PSM. Each team should be required to identify potential performance measures for the processes they are developing and the activities these processes manage. Many of the end-of-pipe measures will already exist; these should be critically examined to ensure that they truly measure performance and are not unduly influenced by other factors. For example, the number of accidents in a fleet of road vehicles is almost directly dependent on the number of miles driven; with no improvement in performance, a reduction in miles driven would reduce the number of accidents.

- *Statistical analysis.* Traditionally, statistical analysis is used to monitor manufacturing quality. However, it is equally applicable in the field of PSM and ESH management. By analyzing accident statistics, we have learned a great deal about human failure rates, for example, the susceptibility of shift workers to errors between the hours of two and four in the morning. Many organizations have been able to significantly reduce medical claims by analyzing these claims and identifying trends or problem areas for management attention and also identify where design or operational improvements to equipment are required.

Continuous improvement does not happen accidentally; it requires a deliberate attempt to include it in the overall processes. The framework developed must require all the elements of continuous improvement, including a feedback loop to assure that the proposed changes do not introduce any new problems or issues.

5.5. Quality Management Tools

There is a large array of Quality Management tools available to help build new systems. The most commonly used tools are described in the following paragraphs. The tools you actually use may be determined by the Quality Management process that is in place in your organization; it is important

5.5. Quality Management Tools

that every member of the team is familiar with the them and understands how they are applied. Not every team member need be an expert, however, at least one member of each team should have an in-depth understanding of the them.

◆**Hint**
Try to select Quality Management tools that the integration team and other staff in your organization are familiar with; otherwise invest time in training.

Management System Model

The framework for integration requires a common management system model to ensure that the development teams design processes that have the same structure and that they consider all the relevant issues. Many of the Quality Management systems propose standard models (although they may not be described as such). Exhibit 5-3 is a Process Characterization Flow Chart developed for the International Electrotechnical Commission (IEC). Such a flow chart ensures that all relevant issues are covered. Also included, after the diagram, are the definition of terms used by IEC in their model. You will need to develop a Management System Model that reflects your own Quality Management system and corporate culture. Such a model can be considered as a higher level tool for the integration process.

Xoff, Inc. identified the following expected benefits for using standard models, which they call templates:

- translates ESH requirements into management systems on a company-wide basis, with measures for continuous improvement
- reinforces credibility with stockholders, governments, communities, and employees
- provides opportunity for a more efficient assessment process
- identifies needed linkages at interfaces between organizations and other systems
- provides mechanism for company-wide, bottom-up improvement process
- provides capability for rapid improvement and communication
- provides benchmark standard for efficiency activity while maintaining effectiveness

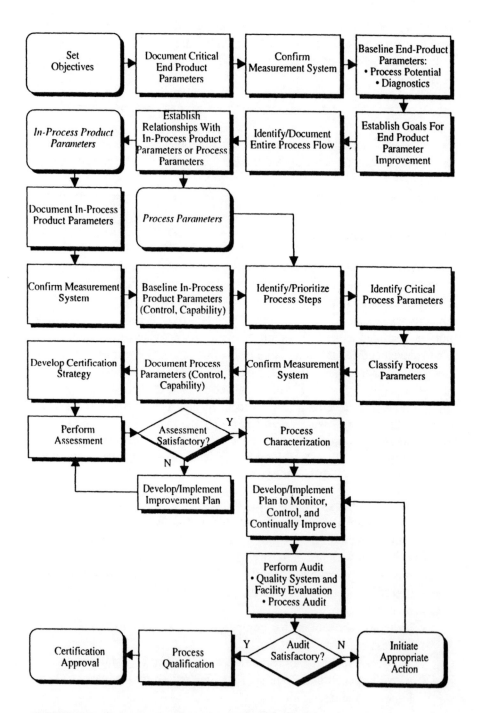

Exhibit 5-3. Example of a Management System Model

> **Exhibit 5-3 (continued).** Example of a Management System Model (Definitions of terms used in diagram)
>
> - *Capability*—The natural variation of the process due to common causes.
> - *Certified Process*—A process that, through demonstration and validation, has been determined to produce product capable of consistently achieving or exceeding customer requirements.
> - *Characteristic*—A distinguishing feature of a process or its output on which variable or attribute data can be collected.
> - *Critical*—Pertaining to that which significantly impacts product quality and/or reliability.
> - *End-Product Parameter*—A parameter that characterizes the product at the finished product stage. This may be piece parts, subassemblies and/or assemblies.
> - *In-Process Product Parameter*—A parameter that characterizes the product prior to the finished product stage.
> - *Parameter*—A measurable characteristic of a product or process.
> - *Process*—The combination of people, equipment, materials, methods, and environment that produce output (product or service). A repeatable sequence of activities with measurable inputs and outputs.
> - *Process Characterization*—The determining of relationships between process parameters and process outputs or product characteristics.
> - *Process Parameter*—A measurable characteristic of a process that impacts product performance but may not be measured on the product.
> - *Process Potential*—The capability of a parameter over a brief period of time.
> - *Product*—The output of a process.
> - *Qualification*—The process of demonstrating whether an entity is capable of meeting or exceeding the specified requirements.

The templates used by Xoff, Inc. all contained the following information:

- Overview
- System Description
    —scope and objectives
    —procedures
    —responsible and accountable resources
    —measurement and verification
    —feedback mechanisms

- References
- Glossary
- Model
  —process descriptions
  —work flow charts

Xoff, Inc. found that it was more cost-effective to integrate globally using their templates, than to have each individual facility or division develop its own approaches.

Benchmarking

Most Quality Management projects conduct some benchmarking studies on selected management processes, programs, or elements. In large organizations, with many different business units, internal benchmarking will be used to identify and learn from best practices elsewhere in the company. External benchmarking may be used where no existing systems exist within your organization or you conclude that you may benefit from examining best practice in other companies. Benchmarking can be considered a more detailed or analytical tool, and might be used in the example given in Exhibit 5-3 to set objectives, identify/prioritize process steps, or help establish goals for end product parameter improvement.

When benchmarking, it is important to recognize that although you have a clear idea of the activity you wish to benchmark others may approach the same activity quite differently and may not recognize what you are looking for. Benchmarking is most successful when focused as narrowly as possible. For example, asking about management of ESH will produce a lot of data, much of it of no use to you, but a request focused on how equipment design changes are approved will provide very specific information. One chemical company tried to benchmark *management of change* among a group of competitors. This study predated PSM and they were given information on regulatory tracking, business strategy, organizational change management, and a little on equipment design and operating change!

◆**Hint**
- Focus the benchmarking efforts on areas where current practices are weak, and on areas where additional information is needed on the potential benefits of integration.
- Ask about lessons learned and pitfalls avoided, not just how the process currently works.

5.5. Quality Management Tools

Particularly when benchmarking competitors or other third parties, it is generally helpful to offer information on your own programs. Today some companies are being asked almost daily to take part in benchmarking studies and they are questioning the benefits they receive. If you offer to share information from your own company and the others being benchmarked, it can help open the door.

The organizations selected for benchmarking should be carefully chosen, since the most useful information may not be gained from your competitors. For each issue try to identify organizations, regardless of industry, that may be using best practice. For example, if you are addressing issues of construction safety, construction companies rather than chemical companies probably have the best practices. After all, they are managing these issues all the time!

Fishbone Analysis

Fishbone analysis can be used to understand the root causes of various problems or failures. However, when looking at a management process do not be tempted to include issues that should be addressed at the program and element level, or look at organization issues that are beyond your scope. Exhibit 5-4 is an example of a fishbone diagram for training failures; note that issues surrounding specifics of the contents have not been addressed as these were beyond the focus of the analysis. Fishbone diagrams are an analytical tool and can quickly become unwieldy. This usually happens when the scope of the analysis has been defined too broadly. Fishbone analysis might be useful to analyze the problems with the existing process before redesigning it according to the model illustrated in Exhibit 5-3.

Process Mapping

While an extremely useful analytical tool for illustrating all the steps in a process as well as critical interactions, process mapping can be a very frustrating exercise either because the process is too difficult to map or because it is very repetitious—the same flows are repeated department after department. It can be a critical part of Process Characterization, as shown in Exhibit 5-3.

It is also important to recognize that there may be no process to map. You have been given all sorts of information by interviewees who felt obliged to answer your questions, even if they did not really know the answer. Look out

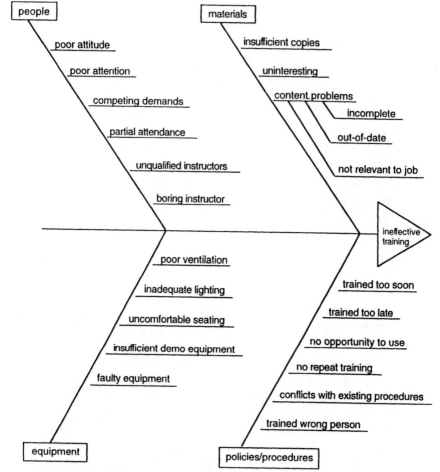

Exhibit 5-4. Sample Fishbone Diagram

for obvious inconsistencies between staff who clearly work together, unexpected answers, suggestions that someone else is responsible for every PSM or ESH issue and so on. Arthur D. Little, Inc. finds that 30 to 50 percent of processes cannot be mapped because they do not really exist.

If you find that there is a very consistent process flow that is repeated in department after department—such as that given in Exhibit 3-4—consider developing generalized maps and then review each department only for differences. The differences can be recorded as notes, by annotating the general diagram or by developing a partial map covering only the areas of differences.

5.6. Converting Informal Systems

In many organizations, the existing PSM and ESH systems may be partially or almost completely informal. It is essential that this informality is eliminated in the integrated system. An informal system exists when there are few written policies, procedures or standards, or written requirements are not enforced or followed. Informal systems almost invariably contain internal inconsistencies which can lead to breakdowns in PSM and ESH management. Different departments and locations will have different systems. Part of the team's work will be to formalize these informal PSM and ESH management during the integration process.

◆**Hint**
Where existing PSM and ESH systems are informal, it is usually better to document or formalize the existing systems (depending on the scope of the system and its criticality) before integrating them, instead of simultaneously formalizing and integrating them.

When faced with informal systems, the team must decide whether to first develop formal unintegrated systems or to jump straight to formal integrated systems. This decision will be based on how well PSM and ESH are actually managed. If the informal systems are delivering a good standard of performance and are well understood by the staff and employees, consider moving directly to formal integrated systems. Conversely, if performance is poor or the existing systems are not well understood, consider the existing programs before moving forward with integration—it may be too large a leap to move directly to integrated programs. Generally, it is advisable to err on the side of caution and formalize existing programs for PSM and ESH before moving forward with integration.

Informal programs inevitably slow down the integration project. Additional steps are introduced into the project and time is required for staff and employees to familiarize themselves with both the individual PSM and ESH systems and the integrated system. However, bear in mind that however informal the existing systems are, it will probably be easy to document them. This is because some documentation will have already been achieved during the assessment of existing systems. Focus most attention on the programs and elements; these will be the building blocks for the integrated system. Interim management arrangements can be used during the switch from existing to integrated systems to manage these programs and elements.

When considering resources for formalizing existing systems, try to avoid using the integration team, as this group should remain focused on integration. Carefully define the scope of the formalization task so that it does not overlap with the integration, or result in unnecessary duplication or rework. Make sure that the formalization team considers the needs of local customers (the workforce, local management, customers, regulators, and the public) and the integration team. The documentation standards and requirements will be set by the integration team.

Reference

1. IEC/TC56, *Process Certification Standard*, Working Draft, August 1993.

6

Testing Implementation Approach

6.1. The Need for Testing

At this point, the ESH systems have been integrated on paper and it is time to prove that the integration is practical, feasible, and beneficial. The effort required to implement new programs across the whole of an organization is enormous. Wasted effort, and rework in particular, cannot be afforded. A pilot implementation on a small scale provides an invaluable learning opportunity. During the pilot study, lessons will be learned and changes to the program made to address these. This chapter describes how to design a pilot study and how to learn as much as possible from the exercise.

6.2. Selecting the Pilot Project

The assessment of existing systems should provide all the information needed to select a candidate for the pilot study. Many of the competing criteria for choosing a pilot study location are discussed in the following paragraphs.

Barriers to Implementation

The barriers to integration identified earlier in Sections 2.5 and 3.2 will influence the selection of the pilot location. Resistance is often very high for pilot studies—most managers would rather have someone else debug a new program before they have to work with it. This resistance is even more likely if the project is viewed as voluntary rather than a regulatory requirement. The resolution of the barriers generally remains the same as for the full program. However, partial or total relief of the costs of implementation may be granted to the pilot study, as many pilot studies are funded centrally with no costs carried locally. This will usually require additional staff to support the pilot study and on-going PSM and ESH activities. If you have residual concerns over a particular location, you might consider a second *unwritten rules of the game* analysis focusing on pilot study issues.

A common concern expressed by managers is that PSM and ESH are costs to their operations that adds little or nothing to customer satisfaction. The use of quality management approaches should ensure that the new systems will be more efficient and so costs will actually fall. It may also be possible to demonstrate that facilities with excellent PSM and ESH records have good performance in other areas. You should look for such examples within your own organization.

Organizational issues may also be raised, as the integrated system may require changes in both the organizational structure and staffing. The selection of the pilot project must address some of these issues to be a credible test, yet too much organizational change will also make it more difficult for the pilot study to be completed quickly and will increase resistance. Therefore it may be useful to pick a location that has more flexibility in its organizational structure–either because it is smaller and individuals have diverse responsibilities or because the key individuals have very secure positions within the organization.

Proactive, Supportive Culture

In every organization, there are departments that are more willing to try new ideas than others. Departments that have recently appointed new managers are likely candidates. These managers may have inherited problems and old loyalties and will welcome a challenging project to bring their new team together. Other departments work in fields where change is commonplace and their staff thrive on this; for example, a specialty chemicals operation may introduce several new products each year and they are familiar with, and relish, the challenge of change. Conversely, departments that are static and have rarely undergone change may resist the idea of a pilot project; for example, a basic chemicals operation focused on producing one product on specification and in planned volume may be

6.2. Selecting the Pilot Project

reluctant to take any risks. Look also for groups that have a history of supporting and buying into quality concepts.

Pilot Should Not Be Too Simplistic

By selecting a location with few PSM or ESH issues to manage (for example a product storage warehouse), or by narrowing the scope of the pilot study (limiting the scope to management-of-change at one facility, say), you can probably guarantee success. But the team will have learned little to help you identify and overcome potential problems with the full project, and will not have done anything to persuade any doubters to change their position. Choosing the toughest department is also inappropriate; although you will learn a great deal, the pilot may be so demanding that the whole project could lose credibility. You should try to select a department that will be representative of the complete project, while at the same time planning for a successful pilot.

Pilot Should Be Able to Measure Improvement

There would be little point in conducting a pilot without being able to measure the improvement achieved. The team should try to find a department where there are good records of PSM and ESH performance, and ideally records that enable the efficiency of the existing management systems to be measured.

These data on improvement will be critical for justifying the full project. If the results are insufficient, there is a possibility that the full project will not be authorized. However, it is unreasonable to expect that the full benefits of integration will be achieved during the life of the pilot project. Some benefits are gained only as the staff become familiar with the new systems. You may be able to use experience of other quality management projects to forecast the full impact of integration.

CCPS has developed guidelines for the measurement of PSM performance that provide useful ideas on measurement. Their approach for developing performance measures integrates Quality Management with a well-tested methodology for developing measures based on mathematical algorithms derived from expert assessment of system performance. The purpose was to identify well-defined, measurable, and practical indicators of performance that are updated on a schedule which makes the performance measure continuously useful. To date, measures have been developed for

Management of Change and Training and Performance. These measures are designed around a number of stages of a program or element's life, such as Design, Development, Installation, Operation, and Testing and Evaluation. The intent of each measure is clearly stated so that the appropriate data are gathered and to ensure that the results are interpreted correctly. The actual measures for the two elements completed so far should be available in 1996.

ESH performance can be measured in a variety of ways:

- *In-process measures.* Possible measures of efficiency (without sacrificing quality) would be the number of hours dedicated to PSM and ESH management, the elapsed (or cycle) time for particular activities (for example, generating routine emissions reports) or the number of PSM and ESH activities undertaken. Other measures to consider are the total number of hours required for different programs and elements (for example, combining Process Hazards Assessments, Environmental Risk Assessment and Job Safety Analysis into a single system could reduce the total number of hours required for the separate activities). If no measures of efficiency exist, it may be possible to construct them by observation and data gathering prior to the pilot study.

- *Compliance with regulations and internal standards.* These requirements usually cover emissions tonnage or concentrations, workplace exposure levels and progress in correcting noncompliances.

- *Measuring waste generated and released.* Virtually every environmental issue is related to the generation of waste. Therefore, quantities of waste generated over time are good environmental performance measures.

- *Ecological impact of operations.* This includes the loss of biodiversity, loss of species, loss of wetlands and restoration of land and waters to environmentally beneficial use. This requires setting a base inventory and then periodically updating it.

- *Human health effects of operations.* This requires assessment of such things as workplace exposures; effects of exposure to air toxics, contaminated drinking water and soil; and exposures during product use, misuse, and disposal.

- *Risk measurement.* Few companies have undertaken the quantitative risk assessments necessary to indicate the level of risk they face for individual operations or types of risk, let alone a more holistic, overall assessment of their full portfolio of risks. However, for those that have, monitoring the absolute change in risk can provide a useful measure. For those companies that do not have risk assessments for existing operations, it is possible to measure the reduction (relative change) in risk achieved by environmental and safety projects. For the pilot project, it may be possible to undertake a risk assessment before and after the pilot project has been installed and operated for some time.

Improvements in safety and environmental performance will come only slowly. Major accidents are rare, even with poorly managed programs, and any reduction in these will be evident only after several years. Areas where early improvements might be expected are spill response and occupational injuries where improved processes should be effective soon after installation. Work place exposure and environmental emissions may improve over a period of several months.

The team should be realistic about the time required to see improvements in end-of-pipe measures; in most cases the pilot project success will be measured on efficiency improvements and other in-process measures alone. In this case it is important to demonstrate that all PSM and ESH issues are being managed. You should consider having a management systems audit (validation) conducted by a group independent of the integration project team. This may be done in conjunction with the next scheduled audit. This may be a corporate or divisional audit function or a consultant engaged specifically for this task.

◆**Hint**
If there are no existing measures of PSM and ESH, do not panic! Use the results of an independent audit of the management systems conducted before and after the pilot study or use the results of existing audits.

6.3. Establish Success (and Failure) Criteria

The results of the pilot study will be most credible if success criteria are established ahead of time. This avoids the temptation to adjust criteria to demonstrate success. Success criteria should be related to the benefits of integration previously identified. Ideally they should be able to provide an indication of the economic benefit achieved. For example, reduction in personnel required to manage PSM and ESH, reduced documentation requirements, reduction in lost time injuries, reduced fugitive losses and, in many instances, improved operational performance. You should also consider what might constitute failure and cause the abandonment or modification of the full project.

◆**Hint**
- Establishing success criteria in advance of the pilot project improves the credibility of the findings and conclusions.
- The team should share your success criteria in advance with senior management, in particular your sponsor and other key staff. Find out whether they agree with the criteria, and if they do not, modify the criteria.

Key measurements relate to the improved use of resources, that is the cost of managing PSM and ESH and maximizing the benefit of all ESH activities. Indicators of this are:

- Changes in the number of steps in management processes
- Total number of processes
- Number of different documents required by PSM and ESH programs and elements
- Number of duplicate or nonessential tasks eliminated
- Number of inadequately controlled risks identified and corrected, along with the appropriate allocation of capital to risk reduction activities
- Time required to incorporate new regulatory requirements into operating procedures, training programs and other processes.

The team should also measure the success of the pilot project itself—compared with the plan how long did each activity take, or how many new processes were developed, or how many programs and elements needed to be developed or upgraded compared with your plan? This will allow you to reconsider the overall project cost estimate and revise it if necessary.

6.4. Communication

As always, communication is critical to the credibility of the pilot study. Communication must include those directly affected by the pilot study and

6.4. Communication

those who will be required to make decisions based on the outcome. You should also remember to provide general communication throughout the organization, as this will ensure that the integration project will not be "forgotten" and take everyone by surprise when the full project is rolled out. The communication program could also help overcome reservations held by some staff by demonstrating that their concerns were successfully addressed or were unfounded.

Generally, the pilot test will be focused on a group small enough for you to conduct a focused communication activity involving everyone who will be affected. Members of the integration team can spend time with individuals and work groups. This should include shift workers who work outside of regular office hours. These communications can be used to help generate an atmosphere of involvement, so keep the meeting relatively informal and ask for input on likely problems and how these might be resolved. Corporate or divisional communications staff can help prepare articles on preparation for the pilot study and report on the results. Communications should cover expected problem areas and subsequently report on how these were overcome and what unexpected problems had to be addressed.

♦Hint
- Regular communications before, during, and after the pilot test ensure that the integration project remains in the minds of key people. Issues you might discuss are:

    —*Before:* Schedule for and expected duration of the pilot project and selection of pilot location (see Attachment 6.1 Sample Pilot Project Advance Communication at the end of this chapter)

    —*During:* Progress against milestones, unexpected benefits identified and outlook

    —*After:* Benefits and costs, lessons learned and changes to overall project

- You can use the communications program to reenforce the themes of integration and avoid the need to reeducate staff once you move to the full project installation.

6.5. Conducting the Pilot

The pilot study is likely to use all the same Quality Management tools that were used for the integration framework development (Chapter 5). However, the limited scope of the pilot study enforces some limitations and compromises—it will not be possible to make changes outside of the department covered by the pilot. Any existing interfaces with other departments must remain the same. This will impact the design phases of the work. For example, Material Safety Data Sheets (MSDS) may currently be prepared locally, in the overall project it might be proposed to develop these centrally. However, for the pilot study it will not be possible develop the central resource so the pilot would have to continue to rely on local resources and there will be no efficiency improvement.

The use of the pilot study to confirm the benefits, costs, methodologies, and integration framework imposes some extra requirements. Additional resources will be required to determine the effectiveness of existing systems and to measure the improvements as early as possible. As mentioned above, the team should consider using outside resources for these activities.

It is important to remember that unexpected benefits may arise from integration. You should actively look for these benefits and document them. It may be possible to improve some of these benefits by small modifications to the plan or integration framework. Such additional work should be undertaken only with the appropriate approvals. Never the less, if any benefits would only be achieved with the integration project, you should include them in your overall statement of benefits. An example of this might be the better allocation of capital to risk reduction efforts when an integrated risk assessment is done–addressing several different types of risk.

◆**Hint**
Keep the pilot project design and installation as similar to the full project as possible. In this way, any lessons learned from the pilot will be most easily transferred to the full project.

You can also use the pilot study to provide a training ground for the integration team, with care to avoid overwhelming the subject department with the whole integration team. It is unlikely that you can find realistic opportunities to involve the whole team directly. But you may be able to have those not directly involved "shadow" the pilot team by meeting with them regularly and discussing issues and their resolution.

If you have chosen your pilot carefully, you can offer the option of reverting to the old way of doing things when the pilot is complete. Not only are you unlikely to be taken up on your offer, you may turn the pilot facility into a real advocate for the new, integrated approach.

6.6. Identifying and Correcting Deficiencies in Integration Plan

The most important objective of the pilot study is testing and improving the integration plan. Deficiencies may be uncovered in the design or installation process. Problems with the installation will usually be evident immediately, as will some design problems; other design problems may not be evident for sometime.

The integration team should review the design and implementation process. This will usually consist of two steps:

- Comparison of the planned resources and timetable with actual
- Team analysis of the plan. This will usually take the form of a meeting, or series of meetings, where Quality Management techniques are used (see below) to analyze the pilot study.

The review meetings usually identify areas where the greatest gaps between expectations and outcome are perceived and then rank these by the likely improvements that will be achieved. This ranking process usually consists of an initial brainstorming discussion led by a facilitator with someone recording the key points on a flip-chart. The brainstorming session is followed by multivoting to select the areas for analysis. In multivoting everyone in the meeting is given a number of votes (usually somewhere between three and ten) to use as they wish; all the votes could be applied to one issue, one vote to each of several areas or any other combination. Beginning with the highest priority areas for improvement, the team will review the process used by applying techniques such as flowcharting.

Some deficiencies in the integrated systems may be seen at once; others will be visible only after project resources have been withdrawn and the system is self-supporting. It is important to examine the integration straight after installation and a few months later. These reviews should draw on the results of the independent audits and should involve staff from the pilot location. The process will be similar to that used for analysis of the design and implementation processes.

◆Hint
Keep in regular contact with key managers at the pilot location throughout the installation of the full project. The need for some modifications may take some time to become apparent. By keeping in regular contact, you will learn of these as early as possible.

It is possible that the later analysis will be conducted while the main project is already underway. If this is the case, you should allow time and resources for adjustment of the main project following the analysis.

Remember that the objectives of the pilot are to refine the integration methodology and to estimate the expected costs of the project and benefits that integration will provide.

Reference

1. Stephen Poltorzycki, *Environmental Performance Measurement*, Arthur D. Little, Inc. Perspectives Series, 1992.

Attachment 6.1. Sample Pilot Project Advance Communication

Environmental, Health, and Safety Revolution

The Osprey River facility is about to undergo a revolution! Osprey River has been selected for a pilot study to integrate all aspects of EHS into a single management system. This is the first stage of what will be a company-wide project. The pilot project will be used to test the design and installation strategies and to firm up the costs and benefits of the overall project.

The objective of the pilot is to integrate the existing Process Safety Management (PSM), Environmental Management, Occupational Safety and Industrial Hygiene systems into a single management system. The integration will be based on the ISO 9000 system adopted throughout the company.

Sandy Schreiver, manager of the integration project, told us that the project is expected to pay for itself in reduced management costs and

Attachment 6.1. Sample Pilot Project Advance Communication

improved ESH performance within 18 months. He hopes to confirm these projections through the pilot project. Examples of reduced costs include: fewer audits (we presently conduct four separate audits which will be replaced by one overall audit), less paper work for projects and maintenance work, fewer design and operability studies (just as for audits we currently have separate studies for each discipline) and better response to new regulations or internal changes.

The integration team recently completed a survey of existing PSM and ESH management systems within the company. Sandy told us that several important gaps in our existing systems were identified, and the integrated system will close these gaps. For example, most utility plants have been left out of the corporate environmental auditing program, and none of the plastics compounding plants have any form of PSM (they fall outside of the OSHA requirements, but corporate policy requires them to have a PSM system).

Al Bickman, Osprey River site manager, expects the work on the pilot study to start next month and be completed within 12 weeks. He has established a local team of managers, engineers, and operators to work with the integration team. Most of this team will be committed for up to half of their time. The team leader will be dedicated to the project.

Al is convinced that Osprey River will benefit from the integration. As a small site, most ESH issues are managed part-time by staff who are easily diverted by their other responsibilities. He expects the new systems to reduce the time required to manage ESH, while at the same time providing systems that will manage more effectively.

Frank Ragon, company chairman, recently visited the site and met with Sandy and Al. Frank emphasized his interest in improving ESH performance. "Our corporate ESH performance, while acceptable, has failed to improve in the last three years", Frank stated, "I expect this integration project to help us save much of the $10 million accidents and other claims are costing us each year."

7

Tracking Progress and Measuring Performance

7.1. The Need for Tracking and Measurement

Tracking performance is a vital part of any management system, and many management experts maintain that it is impossible to manage effectively without measurement. Representative measurement systems are an integral part of quality management systems. Measurement should examine both the PSM and ESH performance achieved and the efficiency of the management system in order to measure the cost/benefit of the integrated management system.

In the example discussed in Section 4.2, the company found that by integrating its accident/incident investigations across its process safety and occupational safety activities, it has saved one year of duplicate time to date by avoiding conflicting models and coordinating training efforts. This represents a 50 percent reduction in effort.

Some measures of PSM and ESH performance are easy to identify, establish and track. These include accident rates, effluent tonnages and composition and number of days lost to illness. Almost all of these traditional performance measures are end-of-pipe; that is, they measure the output of the management system and allow corrective action only after a failure has occurred. The ideal measurement system identifies potential problems ahead of actual failure allowing corrective action to be taken. This requires using techniques such as audits and hazard assessments.

By identifying the potential sources of failures, it is possible to develop controls to address those hazards. These controls might be passive physical items (e.g., dikes, walls, vents), active physical systems (e.g., fire suppression, pressure limiters, temperature controls), or administrative procedures.

A hazard assessment also provides a means of documenting the need for the various controls and the evaluation of the sufficiency of each control.

Measures of the efficiency and cost of PSM and ESH management are needed so that the overall cost/benefit of the system can be judged and improvement opportunities identified. Of course, PSM and ESH management is designed to avoid costs such as fines, capital damage, business interruption, clean-up costs and insurance premiums. Since there is no direct revenue or product stream that can be measured, surrogates for this must be sought. One example of a surrogate measure is the efficiency of the management system, which can be measured via the resources used, timeliness of key activities and speed of response.

◆ **Hint**
- Develop measures that capture end-of-pipe performance, in-process performance and efficiency and cost of PSM and ESH management.
- Select measures that address the benefits expected from the integration project, such as reduced cycle times or number of hours spent on PSM and ESH management.

There are two ways in which the measurement system is likely to be used. The first is to compare the baseline before integration to whatever progress has been made at a given point in time. The second is as a longer term measure of performance of the integrated systems.

7.2. Capture Early Successes

The integration program was most likely supported by senior management because they saw opportunities to improve performance; that is, better end-of-pipe performance and more efficient management. In most cases, improvements in end-of-pipe performance will take a considerable time to become evident. Measuring efficiency improvements and leading indicators of improved end-of-pipe performance must start as early as possible.

The evaluation of existing systems and the most recent audit reports provide a baseline for efficiency and cost performance from which any improvements can be measured. Historical performance will provide a baseline for end-of-pipe improvements. The measurement system itself will measure improvement; the initial record of performance provides a baseline

7.2. Capture Early Successes

against which subsequent performance can be compared. Exhibit 7-1 provides some possible performance measures in four different categories:

- *End-of-pipe* measures that reflect the overall effectiveness of the integrated management system.
- *In-process* measures that provide early indication of potential breakdowns which could lead to accidents or other unwanted events.
- *Efficiency and cost* measures that track management systems performance.
- *Incentive measures* that provide short-term goals for individuals or groups.

Some of these measures are suitable for routine weekly or monthly reporting, others are more difficult to quantify and might be developed only annually. Examples of nonroutine measures are: tonnage of polluting materials released into the environment, ecological impact of operations, audit findings, hazard assessment findings, risk assessment findings and all of the cost and efficiency measures.

◆Hint
Start measuring the areas that are being used to justify the integration project as early as possible. This will help retain on-going support for the project.

Which measures you actually use will partly depend on the objectives of your integration program. You should be checking that the expected benefits of integration are being achieved. If improved efficiency was a key objective, it will be important to develop and collect measures of efficiency as early as possible. When selecting measures of efficiency, you should make sure you have baseline data from before the implementation of the integration project. If you don't have such baseline data, the full benefits of the project will never become apparent.

End-of-pipe measures continue to be vitally important. The largest PSM and ESH management costs are accident and incident related. If you reduce the costs of managing PSM and ESH, yet accident and incident rates rise beyond any normal statistical variation, the new system is costing the company more. Near misses are a leading indicator for accidents and incidents and should not be neglected.

Exhibit 7-1. Possible Performance Measures for the Integrated System

End-of-pipe measures	Fatal accident rate Lost-time injury rate Capital cost of accidents Number of plant/community evacuations Cost of business interruption Cost of workers compensation claims Number of hazardous material spills (in excess of a threshold) Tonnage of hazardous material spilled Tonnage of air, water, liquid and solid effluent Tonnage of polluting materials released into the environment Employee exposure monitoring Number of work related sickness claims Number of regulatory citations and fines Ecological impact of operations (loss or restoration of biodiversity, species, habitats) Number of staff whose training is up to date (as a percentage of total) Number of staff requiring recertification because qualification has been allowed to lapse
In-process measures	Audit findings or scores Number of near miss incidents Hazard assessment findings Risk assessment results Number of in-service equipment failures Number of alarms and trips found out-of-order when tested Number of safety devices found to be nonfunctional
Efficiency and cost measures	Hours dedicated to PSM and ESH management Number of dedicated PSM and ESH staff (local and corporate) Number of management processes Number of programs and elements Number of audits conducted and time expended Number of hazard assessments conducted Number of ESH studies conducted on a project Number of previously uncontrolled risks identified and corrected Time required for internal approval of changes Time required to introduce new regulatory or corporate policy Number of workers compensation claims Insurance premiums
Incentive measures	Work hours since last lost-time injury Number of days since last accidental release of hazardous material Number of days below plan emission levels Annual objectives or goals

7.3. Measures to Consider

Incentive measures are generally used to draw attention to PSM and ESH performance issues that are, to some extent, controlled by worker behavior or management attention. These measures are frequently used for distributing incentive awards or at-risk pay when particular targets are met, for example 1 million hours without an injury or a target number of changes processed correctly. Whether your company uses incentive measures and awards is a matter of corporate practice.

Undoubtedly there will already be several established performance measures; some of these may be replaced with new measures. It can be important to continue these established measures, at least for a short time. This is because most people will be familiar with these measures and may need time to become comfortable with any new measures.

◆**Hint**
- Retain existing measures for the purposes of comparison and helping staff familiarize themselves with the new measures.
- Only retain existing measures beyond a few months where they are genuinely useful measures of performance.

7.3. Measures to Consider

Exhibit 7-1 listed a number of possible performance measures. Some of these are commonly used and need no explanation; others may be unfamiliar and are described in more detail below. The direction of goodness is not the same for all measurements, so use care in interpreting trends. For some measures, repetitive findings are a particular concern.

Cost of business interruption. The cost of business interruption is significantly influenced by various commercial factors. For example, if you are not producing at full capacity it may be possible to make up lost production at other facilities or when the plant comes back into operation. The business interruption costs will only be the additional costs of production, transportation and any lost sales. As these conditions can vary from month to month it may be more appropriate to adopt a standard method of calculating business interruption costs for PSM and ESH purposes. One commonly adopted approach is to estimate the tonnage of production lost and calculate the cost as being the difference between the sale price and all manufacturing, storage and transportation costs.

Tonnage of air emissions, water emissions and liquid and solid effluent and *tonnage of hazardous materials released into the environment.* These two measures are related to one another. However, the first measure relates the total effluent, including nonpolluting materials. The second measure looks only at the tonnage of hazardous materials contained in the total effluent. Both measures can be important indicators. For example, for solid waste it is important to know the total volume of material for disposal and different upstream treatment techniques may affect the total volume. However, for ozone depleting chemicals, only the quantity of these gases is important and other components such as water vapor may be irrelevant.

Ecological impact of operations provides an indication of the damage, or benefit, you are providing to the environment. Some aspects of this are easy to measure; if you have restoration programs, you can measure their progress, spills of hazardous material may have a immediately measurable impact (number of fish killed, acreage of habitat destroyed, etc.). Others are more difficult; for example, what is the effect of emitting 10 tons of sulfur dioxide? You may need to find surrogates for these such as making assumptions about the number of trees that would be damaged by 10 tons of sulfur dioxide. Finally, in order to place all of this on a equal basis you will need to assign a monetary value to the damage. This can be done by estimating the costs of restoration– such as the cost of restocking fish and the cost of planting and growing new trees. This measure is so complex and difficult to estimate that it is rarely used except during the preparation of environmental impact assessments.

Audit findings or scores are useful periodic measures of improvement in PSM and ESH management. Improvement in performance between audits is generally easy to spot. Audits may not be conducted annually at each facility; however, if a sufficient sample of facilities is audited each year it can provide a useful indicator of overall change across the company. Of course, the audit results are most relevant to the individual facility managers who can develop action plans for improvement based on the results.

Hazard assessment findings are similar to audits, in that they are periodic measures. Hazard assessments might include Hazard and Operability (HAZOP) studies, What-If/Checklists, Fault Tree Analyses, or other techniques. The results of these studies will indicate what types of hazards exist and whether or not these hazards are sufficiently controlled.

7.3. Measures to Consider

Risk assessment results can be used to provide an overall estimate of the environmental, health and safety risk faced by a facility or the whole company. Such assessments place all risks in the same context and allow rational decisions on where investment or management intervention will have greatest impact (biggest bang for the buck). The initial cost of such assessments can be high and the assessment must be updated to reflect changes in design or operation. However, such assessments ensure management avoids creating a new problem while solving another. For example, a one case a vapor recovery system was installed on a tank farm to prevent emissions of hydrocarbon. Unfortunately, the system dramatically increased the risk of catastrophic tank failure due to overpressure.

Number of in-service equipment failures is an early indicator of potential problems. Ideally, maintenance programs ensure that equipment does not fail while in service. Such unplanned failures might, in other circumstances, have resulted in an accident or incident. Tracking in-service failures provides a useful indicator of changes in performance. Monitoring in-service failures led in one case to a recognition that a new additive was damaging valve seals with an associated increase in fugitive losses to the environment. In another situation, such monitoring identified that an increase in the severity of operation was dramatically shortening the tube life in furnaces.

Number of alarms and trips found out-of-order when tested. Alarms and trips are important parts of the control system that prevents accidents and incidents. If these are not functioning, one layer of safeguard is missing. These items are usually subject to regular testing. Recording the number of malfunctions provides an indication of how well these safeguards are being maintained. Trends can be indicative of some change in maintenance programs or equipment that is nearing the end of its useful life.

Number of safety devices found to be nonfunctional. Many processing facilities have high temperature and high pressure relief devices and other forms of protection such as overspeed systems. Frequently these cannot be tested in service and must be removed, often during shutdowns, for servicing. It is good practice to test these devices before cleaning or maintenance to find out whether they are functioning and where possible identify the cause of any problem. As for alarms and trips, monitoring trends can help identify problems, such as gum formation or corrosion, that may have resulted from a change in the operation or old age.

Hours dedicated to PSM and ESH management. This estimate should include not only full-time specialist staff, but also hours expended by staff who spend part of the time working on PSM and ESH issues. In many companies, this will be almost impossible to estimate as no records of how nonspecialist staff use their time is kept. It may be appropriate for the total number of hours to increase if there are new regulations to address.

Number of management processes and number of programs and elements. These measures provide some indication of the diversity of effort required to manage PSM and ESH. Information on these can be gathered during management system audits. If the number of processes, programs or elements change, the drivers for this need to be understood. Most commonly, a new regulatory or internal requirement is likely to result in the addition of a process, program or element. Generally, more efficient management will be achieved by modifying existing processes rather than adding new ones. These measures will provide a useful comparison between the current management system and the integrated system.

Number of audits conducted. This measure is most useful when compared to the planned number of audits and the number of findings per audit. These data allow the number and frequency of audits to be varied, for example, if the number of findings per audit is low the number of audits might be reduced. This measure can also be used to compare to the number of audits that were required before the systems were integrated.

Number of hazard assessments conducted is most useful when compared with the plan and the number of work orders that may have required a hazard assessment. It provides a confirmation that hazard assessments are being conducting appropriately.

Number of independent environmental, safety or health studies conducted on a project, if it is tracked from project to project, can provide an indication of improving efficiency.

Number of previously uncontrolled risks identified and corrected. For the integration project, this data will be relatively easy to collect and will provide an indication of the improved coverage provided by the integrated systems. On an ongoing basis, any new risks should be categorized as being due to changes in design or operation of facilities, or due to oversight. Oversight omissions should be investigated to see if the integrated system needs to be modified.

7.4. Selection and Timing of Measures

Time required for internal approval of changes is a good indicator of how well the integrated system is working. One expected benefit of integration would be a reduction in the number of different PSM and ESH approvals required (where appropriate), as well as a simpler approval process. Generally, the approval process is initiated via project or work order documentation and the signed approval will be dated. Records of each approval will be retained centrally and could be analyzed annually, say, to assess performance. Any changes in performance should be evaluated and followed up.

Number of days since last accidental release of hazardous material. This measure distinguishes between routine emissions (such as from storage tank vents, or low pressure steam discharges) and accidental emissions resulting from maloperation or breakdown. Events that might count would be safety valve releases, accidental releases into inappropriate drainage systems and unconfined spills during maintenance.

Number of days below plan emission levels. This measure generally runs through the plan period used by the organization, or as a rolling total over the last twelve months. This measure encourages analysis of emissions and their sources and can be used to help operators better understand how their actions influence emissions.

7.4. Selection and Timing of Measures

As mentioned earlier some measures will be chosen because improvements in these areas were part of the project justification. It is most likely that these will be efficiency measures. Calculation of these measures generally requires analysis of data or specific data collection exercises. There is a relatively high cost associated with preparing these measures so they should be used prudently. In choosing efficiency measures, you should use only those where you have comparative data about the current management systems. For example, if there is no information on the number of hours dedicated to PSM and ESH, don't use this to try to demonstrate the improvement in efficiency.

For many of the efficiency measures, the relevant data is collected during audits or by analysis of records. It is likely that your company has an internal reporting cycle with which you should coordinate your measurement efforts. However, it may not be possible to change a routine auditing schedule just to meet your needs, so you should consider the audit schedule while selecting your measures.

Although it can be fascinating to collect as much data on performance as possible, there will be a diminishing rate of return. Instead, try to focus on the areas where the greatest opportunity for improvement existed, as were identified during your initial assessment.

◆**Hint**
- Start with as small a number of published measures as possible. You may collect data on many more, but introduce these judiciously; otherwise you may confuse rather than inform.
- Some measures may never be widely communicated, and do not need to be, as they are useful to only a limited number of people to help provide insight into changes in performance.

When selecting in-process measures, try to use measures for which data are already available. For example, avoid using in-service failure data unless the maintenance systems can make this information readily available. These measures will be used to identify potential problems and correct them as early as possible. During the development of the integrated systems, data that will be available for in-process measurement should be identified and measures developed. These measures are most likely to be calculated annually as the volume of data required to provide useful data will be collected only over relatively long periods of time.

End-of-pipe measures are usually reported monthly as part of normal reporting routines. You should consider whether this is an appropriate frequency. Major accidents or incidents will be brought to the attention of the appropriate managers as a normal course of events, and do not occur as often as monthly. Some data may not be statistically reliable when reported over as short a time scale as a month, while other data may be reliable. If you are obliged to report month by month performance, you should consider showing rolling data over a 12- or 24-month period. These rolling data will provide useful trend data. The statistical significance of any variations can also be determined if desired. A sample monthly report is given at the end of this chapter in Attachment 7.1.

7.5. Customer Feedback

It is important to find out how well the new systems are meeting your customers needs. This data gathering should be carefully designed to provide a genuine opportunity to provide feedback in a constructive fashion. It is likely that your company conducts internal feedback surveys from time to time, you should contact those responsible for these and learn from their experiences.

The two most common approaches to seeking feedback are written surveys and group meetings. In either case, you will probably need some specialist help in designing and conducting the survey. It is important that the questions are written in an unbiased fashion that will generate an honest response. If you are gathering data on how well colleagues or managers are performing their responsibilities, you should consider using an independent group to receive and analyze the responses.

You may be able to collect a great deal of useful input via personal conversations about the new systems. This avoids the more costly survey approach, but is also more likely to be biased. Generally, any voluntary opinions are more likely to be negative; human nature is such that complaints are more common than praise. (When did you last thank your bank for improved service? When did you last complain to them?). Ideally, you should set aside a period when you simply spend time meeting with other staff and discussing the new systems. If you are given, or hear of, complaints informally, check them out by asking others their opinion.

◆**Hint**
Don't rely on informal customer feedback. Deliberately seek out feedback using a structured and professionally designed survey tool.

7.6. Improving Performance

Improving performance is the objective of measurement. Thus measures should be capable of analysis to help identify where there may be weaknesses that can be corrected. For example, injury data may be published division by division. If one division has deteriorating performance, it will be important to find out whether the deterioration is common across the whole division or specific to one location.

Attachment 7.1. Sample Monthly Report

The sample monthly report presented on the next three pages is made up of different reporting formats; an actual report would probably use a more consistent presentation format. The report is a monthly corporate level report; it assumes the reader is familiar with the format and provides little explanation or interpretation. Each division or facility would have its own report which would cover local issues in more detail.

Sample Monthly Report

24 month rolling Lost-time injury rate

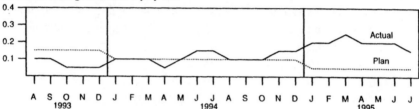

Daily tonnage of hazardous materials released into the environment (monthly average)

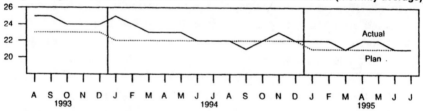

Number of regulatory citations and fines per month

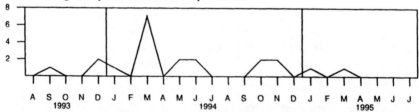

Event summary

Two lost time injuries were incurred at the Wyoming Bay facility when a stack of palletized valves in a warehouse were knocked over by a fork-lift truck.

The upgraded personal protective equipment program at the former Acquired Inc. facilities is beginning to impact the injury rate.

There were no unplanned releases of hazardous materials this month, and daily emissions were lower than plan throughout the month.

Two citations issued in April 1994 have been dismissed on appeal.

Attachment 7.1. Sample Monthly Report

Sample Monthly Report

Individual Facility Incentive Performance

	Osprey River	Wyoming Bay	Poole Harbor	Westing Height	Appleyard	Aberdeen Valley	Billericay Town
Hours since last LTI	251,101	2,567	156,788	2,220	888,195	901,500	150,987
Days since last release	56	124	196	64	153	224	132
Days below plan emissions	196	179	150	180	190	204	65

Number of in-service equipment failures

	A	S	O	N	D	J	F	M	A	M	J	J
Osprey River	5	4	11	7	1	1	1	2	1	2	1	2
Wyoming Bay	7	6	8	5	6	4	6	7	4	4	3	4
Poole Harbor	1		2	1		1		1	1	2		
Westing Height	9	8	11	7	7	6	8	7	7	5	6	7
Appleyard	5	6	7	4	4	4	5	7	3	4	3	6
Aberdeen Valley					6	5	6	7	5	5	6	6
Billericay Town					7	5	7	7	4	5	5	4

Commentary

In-service failures at the Aberdeen Valley, Billericay Town, and Westing Height continue at a high rate. The new maintenance program will be installed by year end.

There have been above plan emissions in the last two months, however this may be due to lower production rates.

Osprey River gas turbine suffered a blade failure and the casing sustained some damage. The cause of this failure is under investigation.

Sample Monthly Report

Integration Project

The integration project has completed the pilot project. The results are currently being analyzed.

Based on preliminary assessment the incentives for integration are now projected to be 10 percent higher than projected in the project justification.

Roll-out of the total project will commence during the fourth quarter. The remaining schedule is shown below.

Efficiency Improvement

The pilot project has reduced the number of corporate and facilities audits by 75%. The separate audits for PSM, environment, industrial hygiene and occupational safety have been replaced by a single audit. The total number of hours has been reduced by 25%.

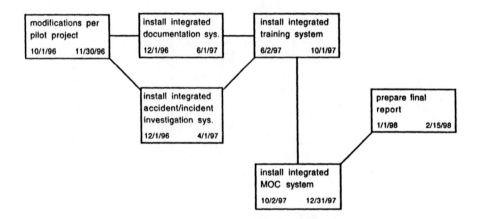

8

Continuous Improvement

8.1. The Need for Continuous Improvement

Integration of EHS elements should produce performance improvements; however, there will always be opportunities for additional (i.e., continuous) improvement. Furthermore, there will need to be an ongoing assessment of the effects of the integration to ensure that there is not a deterioration in performance over time. This might result if reductions in staffing cause overloading of the remaining staff, particularly as new demands on their time arise from regulatory or other changes.

In many organizations, continuous improvement is the justification for introducing Quality Management systems. The business climate demands that organizations respond to change swiftly and efficiently; those that fail to do this will be left behind. Industry found change and improvement increasingly difficult to achieve with traditional approaches to management. Other approaches to change frequently result in continuous correction, where frequent adjustments are required to make the system function as intended.

Continuous improvement includes improving the management of current issues and expanding (or contracting) to manage new issues. Improvement in management of existing issues is most likely to be a better cost/benefit balance. However, in some cases improvement might be more focused, for example, eliminating regulatory citations. Another example of the need for improvement is response to regulatory change; the need to react to these changes can be viewed as being one aspect of continuous improvement.

As described in the previous chapter, Quality Management requires that measures of performance be established. These measures are intended to be used to monitor quality and improvement in performance.

The very design of Quality Management systems is intended to guarantee continuous improvement. Many of the features discussed in the earlier chapters contribute to continuous improvement. This chapter summarizes these features and demonstrates how they relate and combine to provide continuous improvement.

Quality Management approaches encourage a comprehensive review of a system. This comprehensive approach encourages the designers to question every aspect of a system. In this way, incremental improvements are quickly identified and by implementing new approaches dramatic improvement in performance can be achieved.

◆**Hint**
- In most organizations, the driving force behind Quality Management was to achieve continuous improvement. Even if you see other justifications for integration, many of your colleagues will be more interested in continuous improvement.
- Include discussion of continuous improvement in your communications.

Aspects of ISO 9000 that contribute to continuous improvement include:

- *Management responsibility* requires that each program, element or management process is overseen by a manager. This individual should take responsibility for putting improvement initiatives into motion.
- *Auditing the quality system* ensures that performance is regularly reviewed and deficiencies identified and action to address these required.
- *Product verification* requires inspection and testing programs to confirm that "product" meets the quality requirements. In PSM and ESH terms, the product is the management of environmental, health and safety issues and may be difficult to separate from measurement (Chapter 7) and Audit.
- *Nonconformity* evaluations ensure that the root causes of any failures are identified and corrected. This ties in closely with *Corrective Action*.

- *Personnel (training)* ensures all staff know how the systems should work and equips them to identify breakdowns and help identify and correct any underlying problems.
- *Use of statistical methods* requires that performance be analyzed to identify weaknesses and strengths. Weaknesses can be addressed and strong systems used as models for improvement.

The role of each of these in continuous improvement is expanded in the following sections. While these points are relatively generic to any process improvement, there are several continuous improvement opportunities which are unique to integrated PSM and ESH systems:

- *Compliance demonstration* where audits quickly identify noncompliance with regulations and standards, allowing early correction and avoidance of penalties.
- *Easier integration of new requirements* is achieved through a comparison of new requirements with the objectives of the existing integrated system. In this way, those new requirements that are already covered fully or partially are quickly identified. Modifications and additions to the integrated system will be designed and installed using familiar processes.

◆**Hint**
Evaluate your Quality Management system to identify the role each aspect of the system has in achieving continuous improvement.

8.2. Management Responsibility

Management is ultimately responsible for defining the objectives of any activity. In the fields of PSM and ESH, top management might, for example, define the corporate objective as to be "recognized as the leader in PSM and ESH performance in our industry" or "to be in compliance with all pertinent regulations and industry standards."

Part of management responsibility, therefore, is initiating any changes needed to keep performance on target. Within a Quality Management system, the actual process for designing and implementing change will be shared among those who own the relevant programs, elements and management processes. Management is also responsible for ensuring that customer needs are being met, in this case employees, neighbors, owners and regulators.

For example, near-miss incidents in one company had not been reduced significantly despite a new, quality-based system that had been in place for more than a year. The department manager ordered that the near-misses be analyzed more thoroughly. (It should be noted that such proclamations by management must be made with care, as otherwise they may simply serve to discourage accurate reporting.) These assessments revealed improperly maintained instrumentation that was directly responsible for a majority of the near-miss incidents.

8.3. Auditing the Quality System

Audits are designed to uncover successes and failures of a management system, in this case the integrated management system that has been developed for PSM and ESH. Two categories of breakdown are usually evident: one-time problems that are a result of a single error in following the system, and systematic or repetitive issues that are the result of an inherent weakness in the management system.

When looking at one-time problems, it is important to consider whether this could have directly led to an undesired consequence, such as an accident or incident or a noncompliance situation, or if there is another level of safeguard. If it is concluded that the problem or weakness could have directly led to an undesired consequence, the system design should be reviewed to find a way in which a safeguard can be provided. Systematic issues require a redesign of the system to remove the potential for the problem.

As an example, assume that the integrated system has combined several different types of incident reports into one master report which addresses health and safety, environmental and business interruption/property damage impacts. A one-time problem might be the omission of a required signature because someone was out of the office when the report was filed. This would not generally lead to an undesired consequence. A systematic issue might arise if the instructions for the report are unclear or incorrect, such that incidents considered reportable by one or more government agencies are not being reported. This could lead to a fine and necessitates a change in the instructions and training and/or communication about the change.

8.4. Product Verification

For PSM and ESH, product verification can be interpreted as whether or not performance is meeting expectations or established targets. Clearly it is important to decide first whether a target was overly ambitious; if it was, no corrective action may be required. If operating conditions have changed significantly (for example, production rates are low and part of the facility is shut down), you should reassess performance which might have been influenced by this (less equipment in operation will reduce the potential for spills and emissions, but will not necessarily affect ongoing migration of contaminants through the soil, for example). It may be necessary to redesign or adjust measures from time to time to reflect permanent changes in operations and/or regulatory requirements. Targets or goals which are based on number of tasks completed or hours expended are most likely to need such adjustments.

Differences between expectation and reality should be analyzed and the root causes of these identified. When appropriate, changes to the management systems should be developed and installed. For instance, one goal of the integration effort might have been to reduce the number of hours spent in audits by 50 percent. This was initially considered quite reasonable and achievable, because three separate audits (with process safety, occupational health and safety and environmental focuses) were going to be combined into one comprehensive audit. This audit would take longer than any one of the others, but still offered significant opportunities for improved efficiency.

However, tracking the actual hours expended showed that there was only a 20 percent reduction in hours. Was this because the auditors were resistant to their changed responsibilities, or were auditors being selected who did not have the requisite knowledge to cover several different areas such that the teams were larger in size? In fact, closer examination showed that the initial goal was unrealistic, because the audits were not sufficiently comprehensive before the integration effort, and because new state laws had increased the frequency with which such audits were to be conducted. If the root cause had turned out to be that the auditors selected were not appropriately skilled, it might have been necessary to train some auditors or to change the criteria for the selection of an auditor.

8.5. Nonconformity and Corrective Action

There should be a process for addressing and correcting nonconformity. In PSM and ESH, nonconformity could be a failure to meet a compliance standard, a breakdown in a management system or an accident or incident. Such nonconformities will be identified through the measurement system, audits or accident/incident investigations. There needs to be a process for managing corrective actions; the best known example of this is PSM management of change.

The change management process ensures that modifications are thoroughly reviewed to prevent new risks being introduced or the management processes, programs or elements being compromised in any way.

The change management process should also ensure that a root cause analysis has been conducted to make sure the real problem has been identified and corrected. For example, if a pump seal fails it could simply be replaced with an identical seal. However, it may have failed because it was left in service beyond its natural life and the real failure was in the preventative maintenance program that should have replaced it earlier.

Ad hoc or entirely local solutions should be avoided; the system should be fixed rather than placing a bandage to stop the bleeding. The whole process should be revisited to find the most effective way of solving the problem. You might train a few individuals if they are the only ones having problems filling out the incident reports, but if it is a more widespread problem, the instructions or form itself may need to be modified.

Although not normally recommended, in some cases an ad hoc solution may be used because a complete redesign would cost more in effort than the improvement in efficiency or performance would justify. However, where such interim actions are taken it is important to maintain a record so that when the opportunity for a redesign does arise the ad hoc fix will be recognized and a more comprehensive solution developed.

8.6. Personnel (Training)

Training provides an opportunity for encouraging continuous improvement that goes beyond the obvious benefits of improving knowledge and skills. During training sessions, a group of knowledgeable students is focusing their attention on particular PSM or ESH issues. The training sessions can be structured to provide opportunities to discuss existing systems and identify areas where opportunity for improvement exists. During training on new systems, problems can be identified before the system is required to work in practice.

One major corporation uses risk management training to help new managers understand their ESH responsibilities and develop plans for discharging those responsibilities. This program was originally developed during the installation of a newly integrated system, but was found to be so useful it has been continued.

Newly trained staff return to their normal work with a clearer understanding of how PSM and ESH are intended to be managed. They can quickly identify areas where the system has broken down. If staff are trained in the Quality Management techniques (brain storming, process mapping, fishbone analysis) required to analyze and design new systems, they may be able to conduct workshops within their work-groups to improve system components not included into overall integration project redesign.

These improvements are more likely to occur at the program and element level. For example, at one chemical company a group of senior operators took an initiative to redesign the work permit process that reduced the number of forms required, reduced the time required to issue a permit from an average of two hours to 30 minutes and improved the workplace monitoring needed to ensure a safe work environment. The reduction in the time to issue permits increased maintenance productivity by at least 10 percent, by reducing maintenance worker time spent waiting for permits. This more responsive system also reduced the number of instances where work went ahead without the appropriate permit.

The training system itself should also be subject to continuous improvement, particularly in light of the key role it has in ensuring the success of the integration efforts and in achieving overall performance.

8.7. Use of Statistical Methods

The analysis of performance provides a powerful technique for identifying potential for improvement. As discussed in the previous chapter, trends can be spotted and action taken to identify and correct any unwanted development. Additionally, analysis of data can help with the identification of underlying problems. For example, a higher than average number of eye injuries at a particular facility might justify further investigation.

One major oil and chemical company has collected data on the cost of accidents, the equipment involved and the cause of the failure for more than 50 years. These data are analyzed annually to help decide where to focus efforts to reduce losses and/or to modify design standards to prevent recurrence. This analysis also identifies failures of the PSM and ESH management system. These can be compared with the cost of delivering the systems and adjustments made to expenditures to improve the cost/benefit balance. Any such changes must be carefully considered as normal statistical variation may cause you to take unjustified action.

9

Other Quality Management Systems

9.1. Introduction

There are many different management system frameworks that can be used for integrating PSM and ESH. All the concepts presented in this book can be applied to other systems or templates, and the frameworks of different Quality Management systems can even be combined. For example, some one might wish to use the basic elements of ISO 9004 plus a management of change element as their template.

◆Hint
No single Quality Management system is the best choice for all applications; all have their advantages and disadvantages, and are generally based on the same concepts.

Some of the most well known Quality Management approaches are Total Quality Management, the ISO 9000 series, the draft ISO 14000 series, the Malcolm Baldridge National Quality Award criteria, the Deming Prize criteria and the European Quality Award. The ISO 9000 series has been used to illustrate the concepts of this book, and each of the other systems is described briefly below. It is important to remember that the approach suggested in this book is to draw from the framework provided by a Quality Management system, to look at internal (or external) best practices and to integrate these across an organization. This book does not advocate nor discourage certification or pursuit of a quality award.

9.2. Total Quality Management

Total Quality Management (TQM) is not a management system in and of itself. Rather, it is a philosophy of process improvement that incorporates elements of:

- understanding customer needs and relationships
- obtaining data
- using tools
- empowering those who know a process best to fix it
- understanding existing processes
- finding gaps, bottlenecks and inefficiencies within a process
- improving processes or creating them where they did not exist
- developing and using both in-process and results-oriented measures
- looking for new improvement opportunities within a process

TQM has frequently been used to focus on and improve processes in order to meet customer requirements, but not necessarily to look beyond processes for integration opportunities and organizational changes. Thus, it may need to be used in conjunction with another Quality Management system for the purposes of integrating PSM and ESH activities.

9.3. Malcolm Baldridge National Quality Award

The criteria for the Baldridge award in some ways codify the concepts and tenets of TQM into a management system, by describing what an organization that practices TQM should look like. This is not a surprising outcome, as Dr. J.M. Juran was one of the individuals who worked with the National Institute of Standards and Technology on this award. Numerous companies have found that the criteria are useful for self assessment and the identification of obstacles to further continuous improvement, in some cases with additional motivation for change provided by the prize. The criteria provide a systematic description of the management elements that are needed to deliver quality to customers, and cover seven primary areas of evaluation:

- leadership (management support)
- information and analysis
- strategic quality planning
- personnel development and management

- management of process quality (how business processes are managed)
- quality and operational results (measurement)
- customer focus and satisfaction

The criteria themselves are concerned with not only the approach to Quality Management, but also the way in which that approach is implemented and the results that are achieved.

9.4. European Quality Award

This award is similar to the Baldridge award, but also considers the financial success of the company or organization. This additional area of evaluation can alleviate concerns about the real-world viability of companies that win quality awards, but it can also change the focus of the measure to being retrospective. Thus, it may not really consider if a company is positioned for long term success through understanding and meeting its customers needs.

The elements of this system are grouped into nine areas of evaluation:

- leadership
- people management
- policy and strategy
- resources
- processes
- people satisfaction
- customer satisfaction
- impact on society
- business results

9.5. Deming Quality System

The grandfather of Quality Management systems, the Deming quality system tends to be more philosophical than practical in nature. It is less explicit than the Baldridge criteria, and the criteria for the related Deming Prize were designed to be applied by a small, well-trained group of auditors. Some of the concepts which recur throughout Deming's 14 points include:

- leadership
- operating under common objectives and purposes

- relationships
- removing barriers
- reducing reliance on inspection by doing things right
- continuous improvement—plan, do, check, act
- using tools
- on the job training

While Deming's work serves as the foundation of TQM, the Baldridge award and many other quality systems, it requires translation to turn his points into the elements of a management system. His points can be used as a simple checklist, but are hard to measure against or even to interpret consistently—making it difficult to use them as a diagnostic tool.

9.6. ISO 14001

Environmental management systems share the same basic concepts of both PSM and Quality Management systems, however, they recognize a broader range of customers than is typical in Quality Management systems. These might include the public around a facility, society as a whole and evolving societal needs. All of the available environmental management systems are designed to help measure, monitor and assess environmental performance. They offer a high level template or framework that is of use whether or not certification is sought.

The ISO 14000 series is still a draft international standard, but the environmental management systems specification document (ISO 14001) is expected to be approved in mid 1996. The general guidance document is being renumbered from 14000 to 14004 for consistency with the approach used in developing the ISO 9000 series. ISO 14000 has been designed for consistency with ISO 9000 and recognizes that some organizations may choose to use ISO 9000 as their basic management system with a few minor modifications in focus in order to address the concerns unique to ISO 14000. While ISO 14000 does not address occupational health and safety management, it does offer a framework that could readily be extended to cover these and other safety activities, thereby supporting integration efforts.

The draft version of the *Environmental management systems—Specification with guidance for use* document, ISO 14001, is organized into the following sections:

- general
- environmental policy
- planning
- implementation and operation
- checking and corrective action
- management review

The correspondence of this standard to ISO 9001 is also given, and has been summarized in Exhibit 1-1 on page 8 of this book.

References

1. Bradley T. Gale with Robert Chapman Wood, *Managing Customer Value: creating quality and service that customers can see,* The Free Press, New York, NY, 1994.
2. ISO/DIS 14001, Draft International Standard, *Environmental management systems—Specification with guidance for use,* 1995.
3. William Scherkenbach, *The Deming Route to Quality and Productivity: Road Maps and Roadblocks,* CeePress Books, George Washington University, Washington, D.C., 1986.

10

Summary

10.1. Introduction

The key steps to integrate PSM and ESH activities using a Quality Management system are summarized below through a very simplified case study. The approach used in the case study is to indicate the *input, activity and output* of each stage of the integration effort. (Each chapter in these guidelines represents one stage of the integration effort.) For each stage, the input represents the data available and the starting point; the activity describes the main efforts; and the output means the findings or results.

10.2. Case Study

10.2.1. Description

Quality Chemical Company has plants in seven different locations (Osprey River, Wyoming Bay, Poole Harbor, Westing Height, Appleyard, Aberdeen Valley and Billericay Town). Quality Chemical recently conducted a self assessment against the Baldridge award criteria and discovered that they needed stronger policies and better communications throughout their organization, and that their procedures, documentation and training were inconsistent in many areas—including PSM and ESH. Recent reorganizations had also put pressure on various functional areas to increase their efficiency and reduce their resource requirements.

The new corporate vice president of environment, safety, and health (VP, ESH) believed that there were significant benefits to be gained by integrating all the different areas that now reported to him, but was unsure of the magnitude of the potential costs and benefits. Based on his 20 years of experience with Quality Chemical, he believed that the greatest potential benefit would come from integration involving all seven locations, not just at one site. He also wanted to address the full set of ESH programs and elements, as he believed that they had numerous overlaps and redundancies, and was afraid that these would not be uncovered if the scope of the integration effort was too narrow. However, he was not sure that the resulting degree of change would be manageable, so he wanted to approach the integration effort in a very systematic and methodical fashion. This would allow for appropriate decision-making along the way.

The charts below describe the process that the VP, ESH followed to determine the potential benefits, outline the integration plan and implement it. The process is also illustrated through examples in the specific area of documentation systems, which was just one of the many areas that were being integrated at the same time.

10.2.2. Secure Support and Prepare for Implementation

The table opposite outlines the input, activity, and output for Stage 1. It also serves as a summary of the contents of Chapter 2.

The VP, ESH realized that to have just one integration team would slow down progress and would also take a number of key people away from their jobs for an extended period of time. Therefore, he set up a number of integration teams that reported to an overall coordination group. This overall coordination group had the responsibility of ensuring that the teams were efficient in their individual scopes, but did not leave out any of the desired/required programs and elements. There were five teams initially, but this number grew to eight as the areas of interest were better defined and the areas of overlap were identified.

One of the areas in which a team was established from the outset was documentation systems (DS). The experience of the DS team will be used to illustrate the integration process. Comparable efforts were also undertaken by the other integration teams.

The DS team included three ESH representatives who collectively knew the requirements within most of the ESH areas, a corporate documentation specialist and two plant representatives who knew the current engineering documentation systems at their sites. The vice president wanted the DS team to know from the outset that he believed that the area of documentation

10.2. Case Study

Stage 1. Secure Support and Prepare for Implementation

Input	• initiative for integration 　—self audit findings 　—reorganization • initial inventory of existing PSM and ESH programs and elements as given in training manuals, audit protocols and operating procedures • existing Quality Management system at Quality Chemical Company • feedback from interviews on the status of existing processes 　—formal vs. informal 　—adequate vs. broken
Activity	• gathered information on who must buy-in to proposed integration effort • identified potential benefits • put together package and sold the concept • designed the concept 　—established goals and objectives 　—identified resources and leader 　—estimated costs 　—determined primary steps 　—communication/feedback process • evaluated initial inventory of programs and elements to identify potential candidates for grouping/combining • matched grouped programs and elements to QMS to identify synergies and major gaps • developed generic flow diagram for key processes
Output	• corporate support obtained • demonstrated usefulness of QMS for integration • identification of commonalities and integration opportunities • mission statement and goals formulated • integration team formed • scope and approach—a preliminary plan developed

systems was one where increased automation and integration could reduce the head count. Doing so would allow him the freedom to preserve professional positions in other areas.

The DS team spent one session compiling a list of the documentation requirements that they were aware of from their individual experiences. This documentation ranged from vessel details and drawings to accident/incident reporting to MSDSs to safety procedures. Many different departments and functions were involved. Some documentation was kept in hard copy form, while the rest was in electronic form but on several different systems. The exercise showed that approximately 28 of the identified programs and elements had documentation systems in response to requirements. It also showed that these various requirements could probably be reduced to a set of 12 integrated systems by using the Quality Management system that Quality Chemical had developed four years earlier.

For instance, in the area of accident/incident reporting for releases alone, there were reports required by six different federal government programs:

- the National Response Center's Incident Reporting Information System (which in turn fed data to the EPA's Emergency Response Notification System)
- EPA's Accidental Release Information Program
- DOT's Hazardous Materials Incident Reporting System
- DOT's Hazardous Liquid Pipeline Accident Database
- OSHA's Integrated Management Information System
- the Agency for Toxic Substances and Disease Registry's Hazardous Emergency Events Surveillance

These reports were in addition to the internal reporting for OSHA PSM and various local or state requirements. While Quality Chemical had long ago consolidated a number of these programs, there were still four different accident/incident reporting systems—one for external reporting of releases, one for external reporting of injuries and other nonrelease events, one for a corporate-wide database and one for each site itself.

Exhibit 10-1. Potential Benefits of Integrating ESH Documentation Systems

	Number of Documentation Systems		Number of Documentation Training Courses	
	At Present	Plan	At Present	Plan
Accident/incident reporting	4	1	3	1
Equipment/ maint. records (for ESH)	5	2	2	1
Occupational Safety Standards	9	3	3	1 for both
Operational Safety Procedures	7	3	2	
Other	3	3	2	1
TOTAL	28	12	12	4

10.2. Case Study

This potential for consolidation in the area of documentation systems was used as an example to sell the overall integration concept. Documentation was an area of constant complaint and senior management could easily visualize opportunities for improvement in basic resource requirements, as well as secondary effects on resources for training, auditing and other activities. Exhibit 10-1 was a key part of the presentation that the VP, ESH made to senior management.

10.2.3. Assess Existing Management Systems

The table below outlines the input, activity, and output for Stage 2. It also serves as a summary of the contents of Chapter 3.

Stage 2. Assess Existing Management Systems

Input	• initial work on programs/elements and processes from Stage 1 • results from other audits and assessments, including self assessment against Baldridge criteria • manuals and other documentation • information from specialists, obtained via interviews
Activity	• conducted detailed search for all existing programs and elements • looked for overlaps in programs and elements using matrices • gathered data through audits, survey, etc. • used data to assess strengths of management systems • considered unwritten rules that influence both behavior and system performance and looked for ways to address or modify them • evaluated suitability of existing processes and programs/elements for transferring to or expanding for other activities • mapped existing processes via flowcharting or process mapping and evaluated whether processes work and determined who is involved • determined current use of resources—full and part time, dedicated as well as supporting staff • used evaluations as launching point and redesigned processes through quality workshops • updated plan from Stage 1 based on new, in-depth knowledge of both existing situation and desired state
Output	• strengths, weaknesses and overlaps in existing programs and elements recognized • duplications, gaps, nonproductive steps in existing management processes identified • streamlined processes to build integration effort around • updated, detailed plan for piloting and roll out developed

The DS team uncovered four additional sets of documentation systems when they conducted their detailed search of programs and elements. Each documentation system had its own set of standards for recording information, distribution, review and approval and for archiving. Furthermore, the incompatibilities of the various electronic systems were exacerbated by the lack of a corporate hierarchy on preferred computer platforms.

While evaluating the existing systems to look for best practices and the potential for transferring them to other areas, the DS team found that:

- many of the key staff were not adequately trained
- the systems were physically kept in very different areas (some in the engineering department on site, some in corporate ESH and others in human resources—both at the plant and corporate levels)
- people tended to make notes on their copies and developed an unwritten rule of keeping earlier versions; this also meant that a copy was easily obtained when needed in a rush, but that the information might be out of date!
- an in-depth analysis at Westing Height and Appleyard showed redundancy across the two sites and provided an estimate of roughly 8 person-years of effort on documentation systems per year per site, spread among 25 different staff members
- the review at these two sites showed numerous occasions where an out-of-date copy was in use.

After reviewing the documentation systems at all seven locations, one system at Westing Height and two at Appleyard were identified as having the potential for enhancement to serve as integrated processes. Several levels of required approvals were also targeted for reduction, based in large part on changes in the organizational structure that had already occurred. The DS team developed their roll-out plan based on the analyses they had conducted, particularly the in-depth reviews.

10.2.4. Develop a Plan

The table opposite outlines the input, activity and output for Stage 3. It also serves as a summary of the contents of Chapter 4.

The DS team determined that their overall integration approach would revolve around existing local documentation systems, in order to overcome both cultural and resource barriers. The cost outlay for a corporate-wide system was thought to be too prohibitive. Thus, the roll-out would be facility

10.2. Case Study

Stage 3. Develop a Plan

Input	• evaluation of programs and elements against QMS • streamlined management processes • updated plan from Stage 2 • past Quality Management implementation efforts • corporate style and culture

Activity	• determined integration approach and strategy for Quality Chemical Company
	• developed detailed plan
	  —statement of goals
	  —current status of programs, elements and organization
	  —listing of integrated programs and elements that will be developed versus modified and streamlined management processes that will be used as building blocks
	  —tasks to be completed
	  —work products or deliverables
	  —resource estimates—both for integration team and pilot facilities
	  —schedule
	  —expected results—numerical description of changes and improvements
	• verified that all requirements were being addressed
	• determined implementation priorities
	• communicated activities and plans to senior management and others

Output	• integration approach chosen—facility by facility in this case • clear strategy for going forward can be communicated to those impacted for scheduling purposes • enhanced understanding throughout organization as result of following the communication program

by facility, and a schedule could be established. With the proposed approach, at least one of the existing systems would be retained and enhanced at each of the seven sites.

The DS team also decided that they would focus on designing the linkages between existing systems and the establishment of common protocols or standards rather than full fledged redesigns. For example, the data for all the different accident reports would be gathered at once and then the various existing forms would be completed. A totally redesigned approach would have completed all the different forms automatically. In the area of equipment and maintenance records, a common set of standards for maintenance records would be established and an index of all the equipment databases and files would be developed. The records would also be moved to more central locations where possible. This would allow existing systems to be maintained rather than putting all the drawings on AUTOCAD, which remained a long term objective. The index would serve as the linkage between the various systems.

The selected approach was estimated to take 3–6 months for the design stage and one year to implement. The total level of effort was thought to be 2–3 person-years for the first site. Later sites on the schedule would benefit from earlier ones and would be able to see a reduction in these levels of effort to about 1.5–2 person-years, but probably not in the total calendar time required. The total level of effort for all seven locations was estimated to be 13 person-years, and the overall timing was four years, which assumed that roughly two sites would be covered each year.

10.2.5. Design an Integration Framework

The table below outlines the input, activity and output for Stage 4. It also serves as a summary of the contents of Chapter 5.

The integration of documentation systems was a fairly high priority at the outset of the integration effort, and the findings of the individual integration teams did not reduce this priority.

The DS team established their master plan for what the linkages between the different documentation systems should be and what they should look

Stage 4.   Design an Integration Framework

Input	• status of existing processes • status of existing programs and elements • need for local customization • specialized knowledge and expertise from technical staff, operators, etc. • quality tool experience and facilitation skills
Activity	• formalized several informal processes • determined priorities for installation of various QMS elements • developed integrated processes using quality tools and results of Stage 2 • developed integrated programs and elements using quality tools (such as process mapping, flowcharting and fishbone diagrams) and small development teams • addressed continuous improvement • developed installation strategy
Output	• set of fully integrated programs and elements and the processes that support them, developed and documented • strategy for installation determined

10.2. Case Study

like, as well as the overall protocol or standards that would apply to any documentation system at any site. This left only the site-specific customization and implementation to be carried out. Part of the DS team's effort at this stage was also ensuring that their work was consistent with the other integration efforts that were underway. In particular, while the linkages redefined the training needs and vastly reduced the number of different training programs required on documentation, some of these training programs were also able to be combined with other training topics. One example of this was in the area of writing up accident/incident investigations. The need for clarity, specificity, the avoidance of assumptions or inappropriate generalizations, support for conclusions and limiting premature problem solving in writing findings also held true for accident reports, audit findings, process hazard analyses, HAZOPs and other risk assessment reports. A single module in a training course was able to deal with all these situations.

10.2.6. Test Implementation Approach

The table opposite outlines the input, activity and output for Stage 5. It also serves as a summary of the contents of Chapter 6.

In the area of documentation systems, there were several managers who had volunteered their site to be the pilot location. A relatively small location (Poole Harbor) was selected because while it had numerous documentation systems, it was also determined that the overall implementation time would be fairly short. This provided feedback in an expeditious fashion. The remaining managers who had volunteered their sites were used as part of a pilot review team to keep their interest and support active. They actually started to implement the protocols for a few of the documentation systems on their own, in advance of the roll out at their sites. They also became key proponents of the corporate-wide roll out of the overall integration effort, based on their involvement in just this one area.

The DS team found that roughly 95 percent of their design for the system linkages was successful, but that there were a few loose ends in terms of selected drawings that were retained by corporate engineering rather than the individual sites. It had originally been presumed that these drawings would all be relocated to the sites, but that did not prove to be feasible or appropriate. The revised linkages that resulted were added to the master plan as a continuous improvement of the integration effort.

Stage 5.   Test Implementation Approach

Input	• interested facilities from communication sessions and individual discussions • paper designs of new integrated system • selected implementation strategy (facility by facility) • desire for change and improvement

Activity	• selected pilot site
	• selected scale of pilot (several programs and elements or all?)
	• identified and addressed barriers
	• established success and failure criteria
	• communicated with pilot site
	• communicated with others
	• measured existing performance
	• addressed concerns related to transition from old to new
	  —backup for critical items
	  —job security
	  —communications with affected parties
	• trained local staff and operators
	• installed new processes
	• installed new programs and elements
	• documented benefits and problems
	• analyzed results
	• revised plan to incorporate actual data

Output	• support based on actual not potential benefits • highly accurate plan for rest of implementation efforts established • momentum • opportunities for further improvement identified

10.2.7. Remaining Efforts

At this point the VP, ESH and his set of integration teams worked to develop the performance measures that would be used to assess both the success of the overall integration effort and Quality Chemical Company's performance in the areas of ESH and PSM. Some of the measures specific to documentation systems were:

- percent of out-of-date documents
- the number of documentation systems per site
- the number of audit findings in the area of documentation
- annual level of effort required for each site to maintain documentation
- a qualitative rating of the ease of finding material when needed
- the percentage of document reviews that were completed on time

10.2. Case Study

These and other measures were then rolled out along with the integrated processes, programs and elements to the remaining facilities. The set of measures was also designed to indicate areas for continuous improvement.

The vice president had initially felt that the overall integration effort would allow him to retain the existing level of performance and reduce his resource requirements. He also believed that his direct and indirect reports would be able to handle modest changes in existing regulations, but not new regulations. After the test effort was complete, he realized that he would also be able to free up some of the affiliated facility staff to focus on their primary responsibilities in support of operations. After the full integration effort was installed, even greater benefits were realized in terms of reduced overtime hours—which made more time available to manage risks, not just achieve compliance. The overall effort improved the efficiency of resource usage and the effectiveness of programs.

In the area of documentation systems, the specific benefits included:

- increased audit efficiency
- reduced resources at all sites
- a defined document retention policy
- a consistent approval process at all sites for all types of documents (not just ESH)
- a reduced number of copies of most documents
- a decrease in the use of out-of-date information
- increased efficiency in Management of Change activities
- fewer difficulties in gathering PHA pre-work material.

Several specific examples were cited in the VP, ESH's report:

- Errors in instrumentation calibration have been reduced by 90 percent due to the elimination of multiple, outdated copies of calibration procedures/data sheets. An instrument mechanic can now get the current specification from a single, up-to-date source whenever a work order is received. The time wasted in searching multiple files has been eliminated.
- Engineers working on hazard analyses, risk reduction projects and other design changes report that they take 80 percent less time to gather and validate process safety information than before integration occurred. This reduction is attributed to a virtual elimination of piping and instrument walkdowns and searches of local files in Production, Maintenance and Engineering prior to beginning a new

analysis or project. Such efforts are no longer necessary as there is a single, specified location for the latest copy of all drawings, which are verified as correct to the last month. Thus, the engineer can look at these drawings plus the MOC Review File for the most recent changes and know that the set of P&ID's is current and correct.

10.3. Summary

In this book we have shown that many aspects of ESH and PSM management systems are similar, such as auditing, hazard identification and documentation systems. These issues are part of every PSM and ESH program and offer great opportunities for increased efficiency as part of one integrated management system. The case study above illustrated the potential benefits of integrating your ESH and PSM programs, as well as summarized the work process used to accomplish the integration.

The rewards for successful integration are reduced cost of operation, more effective programs and greater recognition of the value of ESH and PSM programs. The lower cost of delivery is achieved by developing management processes with fewer steps and no duplication of effort. Program effectiveness improves by adopting best available practices during the redesign. Finally, programs designed using Quality Management approaches respond faster, at less cost and more effectively to new demands. They also consider the needs of all the stakeholders, increasing the recognition of value by the businesses they serve.

There are numerous other CCPS publications that may be helpful in your integration effort. In particular, the selection of measures to monitor the effectiveness of your integration effort and identify continuous improvement opportunities can be enhanced through the use of the results of a CCPS research and development effort to develop performance measures for CCPS's 12 process safety elements. These results will be described in the CCPS *Guidelines for Developing Performance Measures for Continuous Improvement of Process Safety Management* (scheduled for publication in 1997).

APPENDIX A

Overview of Definitions from ISO 9004: Quality Management and Quality Systems Elements— Guidelines

The basic elements of ISO 9004 are summarized below. Each of these elements appears as a major or minor section of ISO 9004 or ANSI/ASQC Q94. Translations from a quality focus to ESH/PSM are also discussed.

Management Responsibility
This element covers those aspects of the overall management function which are responsible for the effective determination and implementation of a company's quality policy. Translated for the integration effort, quality policy becomes ESH/PSM policy.

Quality System Principles
A quality system comprises the organizational structure, responsibilities, procedures, processes, and resources used to implement quality management, or ESH/PSM management in this case. This part of the standard includes a somewhat detailed discussion of each of the parts of the quality system as well as its documentation.

Auditing the Quality System (Internal)
All parts of the quality (ESH/PSM) system should be reviewed and evaluated on a regular basis via internal audits. This section of the standard

provides guidance on planning, conducting and reporting the findings of audits. Use of the audit findings as part of a periodic review of the effectiveness of the quality (ESH/PSM) system is also described.

Economics—Quality-Related Cost Considerations
The effectiveness of the quality (ESH/PSM) system should be measured to support decisions about improvement efforts, consistent with other financial measures.

Quality in Marketing (Contract Review)
This element addresses the interfaces with the ultimate customer from product specification to customer feedback. For ESH/PSM this may include an understanding of the sophistication of the customers in terms of ESH/PSM issues, and their degree of reliance on the supplier for various information and services.

Quality in Specification and Design (Design Control)
This element covers the translation of customer specifications for the product into the corresponding specifications for materials, products, and processes. Design reviews, product and process acceptance criteria, design change control, and design requalification are all addressed.

Quality in Procurement (Purchasing)
Included in this element are requirements for purchasing documents (specifications, drawings, and purchase orders), selection of suppliers, inspection and control of received material and record-keeping. For ESH/PSM, this element would focus on those aspects of the procurement process which support purchasing according to regulatory requirements, safety standards, risk management controls, etc.

Quality in Production (Process Control)
Issues for consideration under this element include controls or procedures, utilities, raw materials, equipment, and operating environments. Required tests and inspections as well as critical operating parameters, from an ESH/PSM perspective, would be covered here.

Control of Production
This element covers preventive maintenance, management of change (process change control) and verification practices, and frequencies for quality

(ESH/PSM) critical processes or equipment—as may have been identified in hazard analyses.

Material Control and Traceability

Materials—which include chemicals, process equipment, transport containers, protective equipment, and replacement parts—need to conform to quality (ESH/PSM) standards and specifications before being placed in use. Different materials will need varying levels of test and/or inspection as well as controls on their storage and handling.

Product Verification (Inspection and Testing)

This element addresses the quality (adherence to ESH/PSM standards and specifications) of purchased materials and equipment, in-process inspections, and final product verification. Methodologies, frequencies, and sample size or scope of inspections are all expected to be specified.

Control of Measuring and Test Equipment

In order to make sound decisions based on test and measurement data, the equipment used to provide that information must be properly controlled. This section of the standard addresses various control elements, corrective actions, and outside testing.

Nonconformity (Control of Nonconforming Product)

This element describes systems used to identify and address nonconforming materials or products. For ESH/PSM, this might refer to not meeting relevant standards or specifications for materials and equipment or to audit findings of noncompliance.

Corrective Action

The guidance for this element covers responsibilities and authorities, determination of the significance of a quality (ESH/PSM) related problem, root cause investigation, cause-and-effect analysis, preventive actions and additional controls, and change documentation. For ESH/PSM, this element might be interpreted as covering both potential hazards as well as actual problems.

Handling, Storage, Identification, Packaging, Installation, and Delivery

This element covers handling and storage, marking and labeling, packaging, installation (or use) instructions, and delivery (internal and external transportation or distribution), for raw materials and other supplies right through to finished products.

After-Sales Servicing
This element ensures that activities such as installation, operation, handling, and servicing also meet quality (ESH/PSM) standards and practices. Feedback systems on product performance are also addressed.

Quality Documentation and Records
The documentation system should include guidelines regarding the access of records by customers and suppliers, change control procedures, the types of records to be collected, and specifications for indexing, filing, storage, maintenance, retrieval, and disposition.

Quality Records
The types of records needed to demonstrate adherence to requirements and standards as well as effective operation of the quality (ESH/PSM) management system are covered in this element. In addition to many of the quality records listed (e.g., inspection reports, audit reports, operational procedures, and drawings), ESH/PSM requirements might include training records, permits, hazard analyses, audit and other response plans, and accident/incident investigation reports.

Personnel (Training)
Training is addressed in terms of the requirements for all levels and types of personnel, the appropriate motivation of employees and the tracking of performance improvements. Additional ESH/PSM issues may include tracking each employee's training per regulatory requirements, identifying those who require training, maintaining current certifications, ensuring contractors are properly trained, and offering required training within specified time intervals.

Product Safety and Liability
The focus of this element is on the enhancement of product safety and the minimization of product liability through the use of standards, initial tests and evaluations, human factors considerations, and subsequent analyses of problems. For ESH/PSM, safety might be viewed more broadly as safety, health, and environmental protection.

Use of Statistical Methods

This element supports the use of statistical techniques at all stages in the product life cycle, and differs only in areas of focus for quality versus ESH/PSM management systems.

Purchaser-Supplied Product

Purchaser-supplied product is a product or service (such as transportation) owned and provided by the purchaser to the supplier for use in meeting contractual requirements. The quality or ESH/PSM issues around these relationships include ensuring that the product or service received meets the appropriate standards and requirements, that the product is returned in the appropriate condition and that supporting documentation is available as needed.

Bibliography

Karl Albrecht and Lawrence J. Bradford, *The Service Advantage: How to Identify and Fulfill Customer Needs,* Dow Jones-Irwin, Homewood, Illinois, 1990.

Abhay K. Bhushan, "Economic Incentives for Total Quality Environmental Management," IEEE, 1993.

Frank Caropreso (ed.), *Making Total Quality Happen,* Report No. 937, The Conference Board, Inc., New York, N.Y., 1990.

Center for Chemical Process Safety of the American Institute of Chemical Engineers, *Guidelines for Implementing Process Safety Management Systems,* 1993.

Chemical Manufacturers Association, *Questions of Quality, Integrating Process Safety and Total Quality: A Roadmap,Toolguide & Toolbox,* 1995.

Philip B. Crosby, *Quality is Free: The Art of Making Quality Certain,* McGraw-Hill Book Co., New York, NY, 1979.

Philip B. Crosby, *Quality without Tears,* McGraw-Hill Book Co., New York, NY, 1984.

A. M. Dowell, "Regulations: Build a System or Add Layers?" presented at the AIChE Spring National Meeting, Houston, Texas, March 21, 1995.

Bradley T. Gale with Robert Chapman Wood, *Managing Customer Value: creating quality and service that customers can see,* The Free Press, New York, NY, 1994.

David A. Garvin, *Managing Quality: The Strategic and Competitive Edge,* The Free Press, New York, NY, 1988.

H. James Harrington, *The Improvement Process,* McGraw-Hill Book Co., New York, NY, 1987.

Griff Holmes and William Leslie, *Management of Change and Total Quality Management Programs,* Westinghouse Electric Corporation, 1993.

IEC/TC56, *Process Certification Standard,* Working Draft, August 1993.

ISO/DIS 14001, Draft International Standard, *Environmental management systems—Specification with guidance for use,* 1995.

ISO 9000: 1994.

Joseph M. Juran, *Managerial Breakthrough: A New Concept of the Manager's Job,* McGraw-Hill Book Co., New York, N.Y., 1964.

Edward J. Kane, "IBM's quality focus on the business process," *Quality Progress,* April 1986.

Stephen G. Minter, "Quality and Safety: Unocal's Winning Combination," *Occupational Hazards*, October 1991.

John F. Murphy, "Dow Chemical Company's Consolidated Audit," *Proceedings of AIChE 1992 Loss Prevention Symposium*.

Stephen Poltorzycki, *Environmental Performance Measurement*, Arthur D. Little, Inc. Perspectives Series, 1992.

William Scherkenbach, *The Deming Route to Quality and Productivity: Road Maps and Roadblocks*, CeePress Books, George Washington University, Washington, D.C., 1986.

Peter R. Scholtes and Heero Hacquebord, "Beginning the quality transformation, part I; and 6 strategies for beginning the quality transformation, part II," *Quality Progress*, July-August 1988.

Peter Scott-Morgan, *The Unwritten Rules of the Game*, McGraw-Hill, Inc., 1994.

Thomas J. Smith and Thomas L. Larson, "Integrating Quality Management and Hazard Management: A Behavioral Cybernetic Perspective," *Proceedings of the Human Factors Society 35th Annual Meeting*, 1991.

Index

Alarms and trips, number out of order, 127
Assessing support or opposition to integration, 53–56
Assessment of existing management systems, 49–71
  audits, 51
  case study, 153–154
  detailed design considerations, 50–51
  example slides from executive summary, 69–71
  general issues, 50–51
  need for, 49–50
  surveys, 52
Audits
  as assessment approach, 51
  combining, 139
  findings or scores, 126
  number conducted, 128
  of quality systems, 138, 161

Benefits PSM and ESH integration
  continuous improvement, 33, 36
  customer focus and satisfaction, 33
  fewer processes to manage, 36
  higher quality, 34
  less time spent on ESH issues, 36
  lower cost, 34
  measurement, 33, 36
  more effective management of change, 36
  problem solving, 34
  statistical concepts, 33–34
  written standards, 33

Baldridge Quality Award, *see* Malcolm Baldridge National Quality Award
Briefing paragraphs for support of integration of PSM and ESH, 44–47
  cost reduction vs. performance improvement, 45
  funding, 46
  performance targets, 45
  resource allocation, 44
  responsibility for, 45
  timetable, 46
  transition management, 46
  unions and, 46
  why change, 44

Case study on PSM and ESH integration, 149–160
  assessing existing systems, 153–154
  description, 149–150
  designing an integration framework, 156–157
  developing a plan, 154–156
  performance measures, 158–160
  securing support, 150–153
  testing implementation approach, 157–158
Center for Chemical Process Safety (CCPS)
  statement of mission, 1
  other *Guidelines*, 12, 49
Changes, time required for internal, improvement of, 129

Communication
    sample pilot project advance, 118–119
Concerns PSM and ESH integration
    cost reduction, 35
    coverage of all issues, 35
    implementation costs, 35
    span of control, 34–35
Continuous improvement, 135–142
    defined, 135
    management reponsibility for, 137–138
    nonconformity and corrective action, 140
    personnel, 140
    product verification, 139
    statistical measures of, 141–142
Control of production, ISO 9004
    definition, 162
Control of measuring and test equipment,
    ISO 9004 definition, 163
Corrective action, ISO 9004 definition,
    163
Cost of business interruption, 125–126
Customer feedback, 131

Deming Quality System, 145–146
Developing a plan for integration, 73–90
    adjust the preliminary plan, 77–81
    case study, 154–156
    implementation strategy, 81–85

Ecological impact of operations, 126
Economics—quality-related cost considerations, ISO 9004 definition, 162
Efficiency and cost measures, 123
Emissions, tonnage of, 126
Enablers, 53–54
End-of-pipe measures, 123
Environmental safety and health programs
    example, 26
    inventory and assessment of, 56–58
    management responsibility for,
        137–138, 141
    number of independent studies
        conducted, 128
    product verification, 139

Equipment failures, 127

European Quality Award, 145
Existing management systems, assessment
    of, 49–71, 153–154, *see also* Assessment of existing management systems

Fishbone diagrams, 84
Flow charting, 62
    results of, 63–64

Hazard assessments
    findings, 126
    number conducted, 128
Handling, storage, indentification,
    packaging, installation, and delivery,
    ISO 9004 definition, 163–164
Hints
    areas in which continuous improvement
        can be achieved, 37
    assessing management systems, 49
    assessing support or opposition, 54
    continuous improvement, 9, 136, 137
    customer feedback, 131
    developing a list of PSM and ESH
        programs, 22
    developing an implementation strategy, 85
    finding management's concerns with
        integration, 37
    identifying management processes, 28
    identifying who will benefit from
        integration, 19
    measuring improvement, 113, 125, 130
    measuring performance, 122, 123
    obtaining management buy-in, 12
    pilot test, conducting, 116
    pilot test, communication, 115
    pilot test, modifications after, 118
    preparing a cost estimate, 79
    prioritizing programs, 92
    process mapping, 63
    quality management systems, 143
    recasting a plan, communicating, 87
    redesigning management processes
        workshop, 66
    reviewing existing management
        processes, 57
    securing support for PSM and ESH
        integration, 11

Index

Implementing a plan, 81–84
   testing approach, 109–119
Improving performance, 131
   continuously, 135–142
Incentive measures, 123
In-process measures, 123
Integrated systems, developing, 96–97
Integration framework, 91–98
   case study, 156–157
   need for developing, 91–92
   prioritization for installation, 92–92
   project development and installation strategies, 98
Integration of PSM and ESH programs
   benefits and concerns of, 33–37, 152, *see also* Benefits of PSM and ESH integration; concerns of PSM and ESH integration
   case study, 149–160
   commonality within, 22–23, 49–50, 57–58
   defining scope of work and approach, 38–40
   flow diagram of, 29
   identifying who will benefit from, 18–19
   introduction, 1–3
   interview with the chairman, example, 42–44
   mission statement and goals, 37–38
   need for (chart), 3
   preparing a plan, 20–25, 73–90
   redesigning management systems, 65–66
   updating the implementation plan, 67–68
   securing support for, 11–41
Integration plan
   identifying and correcting deficiencies, 117–118
Integration team
   to assess commonalities, 57–58
   selecting, 39
Internal approval of changes, time required for, 129
Interview with the chairman, 42–44
   benefits of integration, 43
   layoffs from integration, 42, 135
   problems from integration, 43
   resources required, 43
   role of chairman, 44
   why integrate, 42
   why ISO 9000, 42
ISO 9000 standards
   comparison of requirements (chart), 8
   groupings for implementation (chart), 84
   matrix of PSM/ESH programs and elements and requirements of, 31
   priority for installation using ISO 9004 (chart), 93–94
   use of, 6
   why use (interview with the chairman), 42
ISO 9004, overview of definitions from, 161–165
   after-sales servicing, 164
   auditing the quality system, 161–162
   control of measuring and test equipment, 163
   control of production, 162–163
   corrective action, 163
   economics—quality-related cost considerations, 162
   handling, storage, indentification, packaging, installation, and delivery, 163–164
   management responsibility, 161
   material control and traceability, 163
   nonconformity, 163
   personnel (training), 164
   product safety and liability, 164
   product verification, 163
   purchaser-supplied product, 165
   quality in marketing, 162
   quality in procurement, 162
   quality in production, 162
   quality in specification and design, 162
   quality records, 164
   quality system principles, 161
   use of statistical methods, 165
ISO 14001, 146–147

Malcolm Baldridge National Quality Award, 144–145

Management process(es)
  continuous improvement, 97–98
  hours dedicated to PSM and ESH, 128
  mapping, 59–62
  number of, 128
  sample analysis of 58–59
Management systems, existing, assessment of, 49–71
Mapping the management processes, 58–65
Material control and traceability, ISO 9004 definition, 163
Measures
  case study, 158–160
  to consider, 125–129
  efficiency and cost, 123
  end-of-pipe, 123
  incentive, 123
  in-process, 123
  performance, 121–134, 158
  selection and timing of, 129–130
  statistical, of performance, 141–142
Monthly report, sample, 132–134
Motivators, 53–54

Need for integration, 1–2
Nonconformity (control of nonconforming product), ISO 9004 definition, 163

Pilot project 109–113
  communication and, 114–115, 118–119
  barriers to implementation, 109–110
  complexity of, 111
  measuring improvement with, 111–113
  supportive culture for, 110–111
Preparing an integration plan, 20–25, *see also* Developing a plan for integration
  case study, 149–151
  initial justification, 21–22
Process mapping, 60–62
  results of, 63–64
Process safety management programs
  comparison of different, 23
  customizing, 27
  example list of elements, 25
  inventory and assessment of, 56–58

management responsibility for, 137–138, 141
number of elements in, 128
number of hours dedicated to, 128, 129, 139
product verification, 139, 163
Product safety and reliability, ISO 9004 definition, 164
Product verification (inspection and testing), ISO 9004 definition, 163
Project development and installation strategies, example, 98
Project phases, 12
Project mission statement, sample, 38
PSM Rule (OSHA), 1
Purchaser-supplied product, ISO 9004 definition, 165

Quality documentation and records, ISO 9004 definition, 164
Quality management programs
  auditing, 138
  bibliographic references (chart), 5
  cost savings achieved using (chart), 4
  Deming Quality System, 145–146
  Dow Chemical Company, 2
  European Quality Award, 145
  with existing programs, 55–55
  inventory and assessment of, 56–58
  ISO 9000, 3, 54
  ISO 14001, 146–147
  Malcolm Baldridge National Quality Award, 144–145
  principles, ISO 9004 definition, 161
  reason for introducing, 135
  and scope/purpose/audience of Guidelines, 3–7
  Total Quality Management, 3, 144
  varieties, 143–147
  Westinghouse Electric Corporation, 2
Quality in marketing, ISO 9004 definition, 162
Quality in procurement, ISO 9004 definition, 162
Quality in production, ISO 9004 definition, 162

Index

Quality records, ISO 9004 definition, 164
Quality in specification and design, ISO 9004 definition, 162
Quality system principles, 161

Recasting the plan, 86–87
Redesigning the management systems, 65–66, 138
Reports, combining, 138
Responsibility, changes in, 139
Risk assessment results, 127
    number of risks corrected, 128

Safety devices, number nonfunctional, 127
Sample plans/project descriptions, 88–90
SARA Title III, 1
Selecting a pilot project, 109–113
Seveso Directive (European Union), 1
Statistical measures of performance, 141–142
Support for integrating PSM and ESH, securing, 11–14
    benefits and concerns covered, 33–37
    dase study, 149–151
    communicating, 16–17
    designing the concept, 15–16
    interview with the chairman, example, 42–47
    of management, 11–12, 26–27
    need for, 11–18
    preparation for (chart), 13–14
    selling the concept (chart), 14–15
    sponsorship roles, 18
    work and approach, defining scope of, 38–40
Surveys, as assessment approach, 52

Testing implementation approach
    case study, 157–158
    communication and, 114–115
    establishing success criteria, 113–114
    need for, 109
    pilot project, 109–113
Total Quality Management, 3, 144
Tracking progress, 121–134
    early success, capturing, 122
    need for, 121–122
Training, 140, 152,
    ISO 9004 definition, 164
Triggers, 53–54

Unwritten rules, 53–55
    assessment of, guide for desinging questions, 55
    example set, 53
Updating benefits and costs, 85–90

Publications Available from the
CENTER FOR CHEMICAL PROCESS SAFETY
of the
AMERICAN INSTITUTE OF CHEMICAL ENGINEERS

CCPS Guidelines Series

Guidelines for Integrating Process Safety Management, Environment, Safety, Health, and Quality
Guidelines for Use of Vapor Cloud Dispersion Models, Second Edition
Guidelines for Evaluating Process Plant Buildings for External Explosions and Fires
Guidelines for Writing Effective Operations and Maintenance Procedures
Guidelines for Chemical Transportation Risk Analysis
Guidelines for Safe Storage and Handling of Reactive Materials
Guidelines for Technical Planning for On-Site Emergencies
Guidelines for Process Safety Documentation
Guidelines for Safe Process Operations and Maintenance
Guidelines for Process Safety Fundamantals in General Plant Operations
Guidelines for Chemical Reactivity Evaluation and Application to Process Design
Tools for Making Acute Risk Decisions with Chemical Process Safety Applications
Guidelines for Preventing Human Error in Process Safety
Guidelines for Evaluating the Characteristics of Vapor Cloud Explosions, Flash Fires, and BLEVEs
Guidelines for Implementing Process Safety Management Systems
Guidelines for Safe Automation of Chemical Processes
Guidelines for Engineering Design for Process Safety
Guidelines for Auditing Process Safety Management Systems
Guidelines for Investigating Chemical Process Incidents
Guidelines for Hazard Evaluation Procedures, Second Edition with Worked Examples
Plant Guidelines for Technical Management of Chemical Process Safety, Revised Edition

Guidelines for Technical Management of Chemical Process Safety
Guidelines for Chemical Process Quantitative Risk Analysis
Guidelines for Process Equipment Reliability Data, with Data Tables
Guidelines for Safe Storage and Handling of High Toxic Hazard Materials
Guidelines for Vapor Release Mitigation

CCPS Concepts Series

Inherently Safer Chemical Processes: A Life-Cycle Approach
Contractor and Client Relations to Assure Process Safety
Understanding Atmospheric Dispersion of Accidental Releases
Expert Systems in Process Safety
Concentration Fluctuations and Averaging Time in Vapor Clouds

Proceedings and Other Publications

Proceedings of the International Conference and Workshop on Process Safety Management and Inherently Safer Processes, 1996
Proceedings of the International Conference and Workshop on Modeling and Mitigating the Consequences of Accidental Releases of Hazardous Materials, 1995.
Proceedings of the International Symposium and Workshop on Safe Chemical Process Automation, 1994
Proceedings of the International Process Safety Management Conference and Workshop, 1993
Proceedings of the International Conference on Hazard Identification and Risk Analysis, Human Factors, and Human Reliability in Process Safety, 1992
Proceedings of the International Conference and Workshop on Modeling and Mitigating the Consequences of Accidental Releases of Hazardous Materials, 1991.
Safety, Health, and Loss Prevention in Chemical Processes: Problems for Undergraduate Engineering Curricula

Printed in the United States
210240BV00003B/18/P